一流本科专业一流本科课程建设系列教材

# 电力专业英语阅读与翻译

主　编　朱永强

参　编　宋　晨　李欣怡　邱　洁　张志涓

机械工业出版社

本书涉及电力行业的多个领域，包括发电厂、输配电系统、电气设备、监测仪表、继电保护、高压与电磁兼容、电力系统稳定、电力电子、电能质量、电力系统自动化、新能源等。

本书共 26 章，分别为电力系统基本概念、电力系统发电、发电站、输电系统与传输线、电力变压器、配电系统和负荷、交流电机、直流电机、开关设备、仪表和传感器、电力系统继电保护、绝缘与接地、过电压和避雷防护、电力系统故障、电力系统稳定、电压调节和无功补偿、电力系统自动化、电力市场、电力系统规划与设计、高压直流输电、灵活交流输电系统技术、电能质量及其改善、电磁场与电磁兼容、可再生能源和分布式发电、综合能源系统、储能。

本书可作为高等学校电气工程及其自动化专业的教学用书，亦可用作电力企业工程技术人员和管理人员学习专业英语的培训教材。

本书提供习题答案和词语翻译汇总资料，请扫描下方二维码查看。

习题答案

词语翻译

### 图书在版编目（CIP）数据

电力专业英语阅读与翻译 / 朱永强主编 . —北京：机械工业出版社，2023.8
一流本科专业一流本科课程建设系列教材
ISBN 978-7-111-75558-6

Ⅰ.①电… Ⅱ.①朱… Ⅲ.①电力工业 – 英语 – 阅读教学 – 高等学校 – 教材 ②电力工业 – 英语 – 翻译 – 高等学校 – 教材　Ⅳ.①TM

中国国家版本馆 CIP 数据核字（2024）第 071874 号

机械工业出版社（北京市百万庄大街 22 号　邮政编码 100037）
策划编辑：王雅新　　　责任编辑：王雅新
责任校对：孙蕾雅　　　封面设计：张　静
责任印制：张　博
北京雁林吉兆印刷有限公司印刷
2024 年 6 月第 1 版第 1 次印刷
184mm×260mm・19.25 印张・484 千字
标准书号：ISBN 978-7-111-75558-6
定价：59.00 元

电话服务　　　　　　　　网络服务
客服电话：010-88361066　　机　工　官　网：www.cmpbook.com
　　　　　010-88379833　　机　工　官　博：weibo.com/cmp1952
　　　　　010-68326294　　金　书　网：www.golden-book.com
封底无防伪标均为盗版　　　机工教育服务网：www.cmpedu.com

# 前　言

本书是编者在多年阅读大量专业文献的基础上，悉心设计、精选细编而成。其特点是内容丰富、覆盖面宽，既有专业知识的英文描述，又有专业知识的合理扩充。书中配有大量的教学辅导环节，虽然以服务本科教学为主要目标，但也力求使研究生、高职院校学生以及电力行业的其他从业人员都可从中获益。

本书涉及电力行业的多个领域，全书共26章，分别为电力系统基本概念、电力系统发电、发电站、输电系统与传输线、电力变压器、配电系统和负荷、交流电机、直流电机、开关设备、仪表和传感器、电力系统继电保护、绝缘与接地、过电压和避雷防护、电力系统故障、电力系统稳定、电压调节和无功补偿、电力系统自动化、电力市场、电力系统规划与设计、高压直流输电、灵活交流输电系统技术、电能质量及其改善、电磁场与电磁兼容、可再生能源和分布式发电、综合能源系统、储能。

各章包含若干部分，各部分有独立的主题，每一部分就是一篇短文。每一章的各部分既有关联，又可拆分。读者可以根据需要，选择适当的章节自学或组织教学活动。

书中提供了丰富的学习辅导内容。正文之后，有生词表、重点词组、难句解说，便于读者阅读理解；还有英文缩写形式列举、相关专业术语总结，帮助读者加深记忆、重点掌握。形式多样的练习题有助于读者检验对本章内容的掌握程度，以便改进提高。相关短语、词组在正文中有下画线标注，难句句尾有*标注。

针对各章节正文中出现的有特点的专业词汇，本书对专业词汇的构词方法进行了总结，边学边总结，有助于扩大词汇量，提高阅读能力。

每3～4章配备一个单元复习，包含重点术语总结、重要的英文缩写列表、专题介绍、补充阅读材料、知识与技巧等内容。帮助读者在做好复习的同时，借助所提供的知识与技巧，实现阅读和翻译水平的提高。

书后附录中给出了全书的词汇表以及总结的英文缩写，便于读者跨章节查阅。

本书的形成过程中，贾玉栋、穆希桢、梁财豪、殷康、李如宁、孙蕾雅、周家欣、刘天泽、周雨桐、李青山等研究生在文字录入、图形整理、章节编辑等方面做了大量的工作，在此一并表示衷心的感谢！同时感谢王欣女士帮助审阅书稿，保证了书稿质量。

本书涉及的内容较广，由于编写水平有限且时间紧迫，书中难免有错误和不完善之处，敬请读者批评指正。

编者

2023年5月

# 目 录

前言

**Chapter 1　Basic Concepts of Electric Power System** ······················· 1

　　Part 1　Basic Construction of Electric Power System ····················· 1
　　Part 2　Typical Topologies of Electric Power System ····················· 2
　　Part 3　Characteristics of Most Power Systems ·························· 2
　　Part 4　The Growth of Electric Power System ···························· 3
　　EXPLANATIONS ···························································· 3
　　EXERCISES ······························································ 8
　　Word-Building (1)　General Introduction 专业词汇和构词方法 ············· 8

**Chapter 2　Electric Power Generation** ······································ 10

　　Part 1　About Power Generation ········································· 10
　　Part 2　Hydropower Generation ········································· 10
　　Part 3　Thermal Generation ············································ 11
　　Part 4　Nuclear Generation ············································ 11
　　Part 5　Application of Various Power Stations ························· 12
　　EXPLANATIONS ··························································· 12
　　EXERCISES ····························································· 17
　　Word-Building (2)　electro-, magneto　电，磁 ·························· 17

**Chapter 3　Power Generating Stations** ······································ 19

　　Part 1　Hydropower Stations ··········································· 19
　　Part 2　Thermal Stations ·············································· 21
　　Part 3　Nuclear Stations ·············································· 22
　　EXPLANATIONS ··························································· 24
　　EXERCISES ····························································· 29
　　Word-Building (3)　hydro-, thermo-　水，热 ···························· 30

**Unit 1** ···································································· 31

　　Review ································································ 31
　　Special Topic (1)　Basic Electric Quantities ························· 32
　　Reading Material —— The One-Line Diagram ······························ 32
　　Knowledge & Skills (1)　Parenthesis ··································· 35

# 目 录

## Chapter 4  Transmission System and Transmission Lines ... 37
- Part 1  General Introduction of the Transmission System ... 37
- Part 2  Considerations for Transmission Lines and Cables ... 37
- Part 3  Circuit Parameters of Transmission Lines and Cables ... 38
- Part 4  Types of Transmission Conductors ... 39
- Part 5  Power Transmission Towers ... 40
- EXPLANATIONS ... 41
- EXERCISES ... 47
- Word-Building (4)  后缀 -or, er; -ance  机，器，装置，元件 ... 47

## Chapter 5  Power Transformers ... 49
- Part 1  Power Transformers Used in Power System ... 49
- Part 2  The Structure of a Transformer ... 49
- Part 3  The Ideal Transformer ... 50
- Part 4  The Equivalent Circuit of a Practical Transformer ... 51
- Part 5  Special Transformers ... 53
- EXPLANATIONS ... 55
- EXERCISES ... 60
- Word-Building (5)  auto-  自动，远程 ... 61

## Chapter 6  Distribution Systems and Loads ... 62
- Part 1  Introduction of Distribution Systems ... 62
- Part 2  Familiar Loads and Their Classification ... 62
- Part 3  Loads Models for Power System Study ... 63
- Part 4  Acquisition of Load Model Parameters ... 65
- EXPLANATIONS ... 67
- EXERCISES ... 71
- Word-Building (6)  re-; im-  重，再；不，非 ... 72

## Unit 2 ... 73
- Review ... 73
- Special Topic (2)  Passive Elements and Impedance ... 74
- Reading Material —— Per-Unit Value ... 74
- Knowledge & Skills (2)  Inversion and Intensive Mood ... 77

## Chapter 7  Alternating-Current Electric Machines ... 79
- Part 1  Components and Principles of Operation of Induction Motors ... 79
- Part 2  Equivalent Circuit and Characteristics of Induction Motors ... 81
- Part 3  Synchronous Generators and Motors ... 82
- Part 4  The Synchronous Motor versus The Induction Motor ... 84
- EXPLANATIONS ... 85
- EXERCISES ... 90

Word-Building (7)　-wise, -ward　方向，方式 ·················· 90

## Chapter 8　Direct-Current Electric Machines ·················· 91

　Part 1　Generating an AC Voltage ·················· 91
　Part 2　Direct-Current Generator ·················· 92
　Part 3　Direct-Current Motors ·················· 93
　Part 4　The Application of DC Machines ·················· 95
　EXPLANATIONS ·················· 95
　EXERCISES ·················· 99
　Word-Building (8)　in-, ir-　否定 2 ·················· 99

## Chapter 9　Switch Devices ·················· 101

　Part 1　Circuit Breakers ·················· 101
　Part 2　Disconnecting Switches ·················· 103
　Part 3　Fuse ·················· 105
　Part 4　Recloser and Sectionalizers ·················· 105
　Part 5　The Selection of Circuit Breakers ·················· 107
　EXPLANATIONS ·················· 107
　EXERCISES ·················· 112
　Word-Building (9)　ab-, a-　否定 3 ·················· 113

## Unit 3 ·················· 114

　Review ·················· 114
　Special Topic(3)　Powers of Ten in Engineering Notation ·················· 115
　Knowledge & Skills (3)　Omitting of Nouns ·················· 116

## Chapter 10　Instruments and Transducers ·················· 119

　Part 1　Importance of Instrumentation ·················· 119
　Part 2　Effects without Instrumentation ·················· 120
　Part 3　Instruments for System Monitoring ·················· 120
　Part 4　Transducers ·················· 121
　EXPLANATIONS ·················· 122
　EXERCISES ·················· 126
　Word-Building (10)　-meter, over-, under　仪表，合成词 ·················· 126

## Chapter 11　Power System Relay Protection ·················· 128

　Part 1　Relay Type ·················· 128
　Part 2　Protection Performance and Protection Area ·················· 129
　Part 3　Composition of Protection System ·················· 130
　EXPLANATIONS ·················· 130
　EXERCISES ·················· 134
　Word-Building (11)　pre-　预，事先；前 ·················· 134

# 目 录

## Chapter 12　Insulation and Grounding ··········· 135

Part 1　Introduction Electrical Insulation ··········· 135
Part 2　Insulation Coordination ··········· 136
Part 3　Grounding ··········· 138
EXPLANATIONS ··········· 139
EXERCISES ··········· 145
Word-Building (12)　前缀 co-; inter- 互，共 ··········· 145

## Chapter 13　Overvoltage and Lightning Shielding ··········· 147

Part 1　Overvoltage and Its Classification ··········· 147
Part 2　Lightning and Its Hazards ··········· 149
Part 3　Lightning-proof Measures ··········· 150
EXPLANATIONS ··········· 151
EXERCISES ··········· 156
Word-Building (13)　non-; dis- 非，不 ··········· 157

## Unit 4 ··········· 158

Review ··········· 158
Special Topic(4)　Symbols for Operation ··········· 159
Knowledge & Skills (4)　The Subjunctive Mood ··········· 159

## Chapter 14　Faults on Power System ··········· 161

Part 1　Faults and Their Damage ··········· 161
Part 2　Distribution System Faults ··········· 161
Part 3　Types of Faults ··········· 162
Part 4　Calculation of Fault Currents ··········· 163
Part 5　Faults Isolation and Recovery ··········· 164
EXPLANATIONS ··········· 165
EXERCISES ··········· 169
Word-Building (14)　magni-, meg(a)-, micro- 巨大，微小 ··········· 169

## Chapter 15　Stability of Power System ··········· 171

Part 1　Basic Concepts and Definitions ··········· 171
Part 2　Classification of Stability ··········· 172
Part 3　Angle Stability ··········· 173
Part 4　Voltage Stability and Collapse ··········· 175
EXPLANATIONS ··········· 176
EXERCISES ··········· 180
Word-Building (15)　-ity, -age －度，－性；总数 ··········· 180

## Chapter 16　Voltage Regulation and Reactive Power Compensation ··········· 182

Part 1　Production and Absorption of Reactive Power ··········· 182

Part 2　Objectives and Methods of Voltage Regulation ……………………… 183
Part 3　Comparative Summary of Alternative Formsof Compensation …………… 184
Part 4　Control by Transformers …………………………………………… 185
EXPLANATIONS ……………………………………………………………… 185
EXERCISES …………………………………………………………………… 188
Word-Building (16)　sub-　低于；子，分，次，亚 ………………………… 189

## Unit 5 …………………………………………………………………………… 190

Review ………………………………………………………………………… 190
Special Topic (5)　Voltage Classes ………………………………………… 190
Knowledge & Skills (5)　Promotion of Adverbial Modifier ………………… 192

## Chapter 17　Automation and Intelligence of Power System ………………… 194

Part 1　Automatic Generation Control (AGC) ……………………………… 194
Part 2　Distribution Automation System (DAS) …………………………… 194
Part 3　Supervisory Control and Data Acquisition (SCADA) ……………… 195
Part 4　Energy Management System (EMS) ……………………………… 196
Part 5　Smart Grid …………………………………………………………… 197
EXPLANATIONS ……………………………………………………………… 197
EXERCISES …………………………………………………………………… 200
Word-Building (17)　-free, -less; -proof　无，免，抗 ……………………… 200

## Chapter 18　Electric Power Market …………………………………………… 201

Part 1　Overview of the Electric Power Market …………………………… 201
Part 2　Physical Power Market and Financial Power Market ……………… 201
Part 3　Power Market Transaction Pricing ………………………………… 202
EXPLANATIONS ……………………………………………………………… 203
EXERCISES …………………………………………………………………… 205

## Chapter 19　Power System Planning and Design …………………………… 207

Part 1　Power System Planning …………………………………………… 207
Part 2　Optimization Method of Power System Planning ………………… 208
Part 3　Content of Power System Design ………………………………… 208
Part 4　Wiring and Operation of Power System …………………………… 209
Part 5　Equipment Selection and Safety Clearance ……………………… 210
EXPLANATIONS ……………………………………………………………… 211
EXERCISES …………………………………………………………………… 214
Word-Building (18)　over-　过度，过分，在……之上，翻转 ……………… 214

## Unit 6 …………………………………………………………………………… 215

Review ………………………………………………………………………… 215
Special Topic (6)　Escape of Words ………………………………………… 216
Knowledge & Skills (6)　Omitting of Verbs ………………………………… 217

# 目 录

## Chapter 20　High-Voltage Direct-Current Transmission … 219
- Part 1　Overview of High-Voltage Direct-Current Transmission … 219
- Part 2　Basic DC Transmission System … 220
- Part 3　HVDC System Configurations and Components … 221
- Part 4　Advantages and Disadvantages of DC Transmission … 223
- EXPLANATIONS … 225
- EXERCISES … 229
- Word-Building (19)　semi-; super-　子，分，次，亚 … 229

## Chapter 21　Flexible AC Transmission System … 230
- Part 1　Concept and Development of FACTS … 230
- Part 2　SVC Family … 230
- Part 3　The Voltage Source Converter … 232
- Part 4　STATCOM … 233
- Part 5　Other Devices Based on VSC … 235
- EXPLANATIONS … 236
- EXERCISES … 240
- Word-Building (20)　bi-, di-; tri-　双，二；三 … 240

## Chapter 22　Power Quality and Its Improvement … 242
- Part 1　Familiar Power Quality Problems … 242
- Part 2　Harmonics and Nonlinear Loads … 243
- Part 3　Voltage Fluctuation and Flicker … 244
- Part 4　Unbalanced Loads … 245
- Part 5　Voltage Sag … 247
- EXPLANATIONS … 247
- EXERCISES … 251
- Word-Building (21)　de-, anti-; counter-　反，抗，逆 … 252

## Chapter 23　Electromagnetic Field and Electromagnetic Compatibility … 253
- Part 1　Overview of Electromagnetic Field … 253
- Part 2　Maxwell's Equations … 253
- Part 3　Electromagnetic Compatibility … 254
- Part 4　Design on EMC in Devices … 256
- EXPLANATIONS … 258
- EXERCISES … 262
- Word-Building (22)　-able; -al　能…的（形容词尾） … 262

## Unit 7 … 264
- Review … 264
- Special Topic(7)　Synthesis of Words … 265
- Knowledge and Skills (7)　Shift of Sentence Structure … 266

## Chapter 24  Renewable Energy Sources and Distributed Generation  269

Part 1  Importance of Renewable Energy Application  269
Part 2  Solar Energy  269
Part 3  Wind Power  270
Part 4  Biomass Energy  270
Part 5  Hydrogen  271
Part 6  Geothermal  271
Part 7  Ocean Energy  272
Part 8  Distributed Energy and Production  272
EXPLANATIONS  274
EXERCISES  277
Word-Building (23)  photo-; bio-  电，磁  278

## Chapter 25  Multi-energy Utilization  280

Part 1  Importance of Multi-energy Utilization  280
Part 2  Energy Internet  280
Part 3  Ubiquitous Electric Internet of Things  281
Part 4  Integrated Energy System  281
Part 5  Energy Hub  282
EXPLANATIONS  282
EXERCISES  285
Word-Building (24)  trans-  越过、超；转移  285

## Chapter 26  Energy Storage System  287

Part 1  Concept and Classification of Energy Storage  287
Part 2  Commonly Used Energy Storage Technology  287
Part 3  The Role of Energy Storage Technology in Power System  289
Part 4  Electric Vehicle  289
EXPLANATIONS  290
EXERCISES  292
Word-Building (25)  dia-;per-  通过，遍及  293

## Unit 8  294

Review  294

## 参考文献  295

# Chapter 1
# Basic Concepts of Electric Power System

## Part 1  Basic Construction of Electric Power System

Electricity is an ideal energy form, which is convenient to deliver and use and is clean without polluting our environment and atmosphere. So it has developed rapidly and has been used widely since it was discovered. The generation, delivery, and consumption of electricity are realized in an integrated system which is called an electric power system or *power system.*

An electric power system consists of three principal divisions: *power generation, power transmission system,* and *power distribution system,* as shown in Figure 1.1.

**Figure 1.1**  The three principal divisions of the power system

All the electricity the consumers take from the power system is generated in a *power plant,* or called *power station,* of some kind.* Power generation is the first stage of the process of the utilization of electric energy.

Transmission systems are the connecting links between the generation plants and the distribution systems and lead to other power systems over interconnections.* A distribution system connects all the individual loads to the transmission lines at substations which perform voltage transformation and switching functions.

Small generating plants located near the load are often connected to the *sub-transmission* or distribution system directly.

Interconnection to neighboring power systems is usually formed at the transmission system level.

The overall system thus consists of multiple generating sources and several layers of *transmission networks.* This provides a high degree of structural redundancy that enables the system to withstand unusual contingencies without service disruption to the consumers.*

## Part 2  Typical Topologies of Electric Power System

The equipment that forms an electric system is arranged depending on the manner in which the load grows in the area and may be rearranged from time to time.*

However, there are certain plans into which a particular system design may be classified. Three types are illustrated: the *radial system*, the *loop system*, and the *network system*.

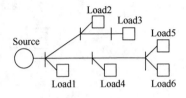

Figure 1.2  A radial system

In a radial system, as shown in Figure 1.2, the lines form a "tree" spreading out from the power source. Opening any line will result in an interruption of power to one or more of the loads.

The loop system is illustrated in Figure 1.3. With this arrangement, all loads may be served even though one line section is removed from service.

Figure 1.4 shows the same loads being served by a network. With this arrangement, each load has two or more circuits over which it is fed.

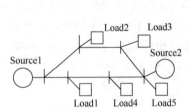

Figure 1.3  A loop system

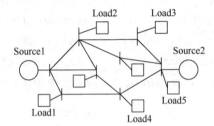

Figure 1.4  A network system

Distribution circuits are commonly designed so that they may be classified as radial or loop circuits. The high-voltage transmission lines of most power systems are arranged as networks. The interconnection of major power systems results in networks made up many of line sections.

## Part 3  Characteristics of Most Power Systems

The first complete electric power system was a direct current (DC) system. The development of alternating current (AC) transmission led to AC electric power systems. The AC system had won out over the DC system for the following reasons:

• Voltage levels can be easily transformed in AC systems, thus providing flexibility for the use of different voltages for generation, transmission, and consumption.

• AC electric machines are much simpler or cheaper than DC electric machines.

Electric power systems vary in size and structural components. However, they all have the same basic characteristics:

• Are comprised of three-phase AC systems operating essentially at a constant voltage. Generation and transmission facilities use three-phase equipment. *Industrial*

# Chapter 1  Basic Concepts of Electric Power System

*loads* are invariably three-phase; single-phase *residential* and *commercial loads* are distributed equally among the phases so as to effectively form a balanced three-phase system.

- Use synchronous machines for the generation of electricity.
- Transmit power over significant distances to consumers spread over a wide area. This requires a transmission system comprised subsystems operating at different voltage levels.

## Part 4  The Growth of Electric Power System

Development of sources of energy to accomplish useful work is the key to industrial progress which is essential to the continual improvement in the standard of living of people everywhere.* The electric power system is one of the tools for converting and transporting energy which is playing an important role in meeting this challenge. The industry, by some standards, is the largest in the world.*

The commercial use of electricity began in the late 1870s when arc lamps were used for lighthouse illumination and street lighting.

The first complete electric power system was built in New York City which began operation in September 1882. This was a DC system. This was the beginning of what would develop into one of the largest industries in the world.

The development of AC systems began in the United States in 1885. The first AC transmission line in the United States was put into operation in 1890. The first transmission lines were single-phase. With the development of polyphase systems, the AC system became even more attractive. Thereafter, the transmission of electric energy by alternating current, especially three-phase alternating current, gradually replaced the DC system.

In the early days of AC power transmission, the operating voltage increased rapidly. For instance, in the United States, in 1890 the first line was operated at 3.3kV. Voltage rose to 100kV in 1907, 150kV in 1913, 220kV in 1923, 244kV in 1926, and 287kV in 1936. In 1953 came the first 345kV line.* The first 500kV line was operating in 1965. Four years later in 1969, the first 765kV line was placed in operation.

Until 1917, electric systems were usually operated as individual units because they started as isolated systems and spread out only gradually to cover the whole country. The demand for large blocks of power and increased reliability suggested the interconnection of neighboring systems. Interconnection has increased to the point where power is exchanged between the systems of different companies as a matter of routine.

## NEW WORDS AND EXPRESSIONS

| | | | |
|---|---|---|---|
| 1. | electric | *adj.* | 电的，用电的，电动的，发电的 |
| 2. | electricity | *n.* | 电，电学，电流 |

| 3. | energy | n. | 能源，能量 |
| 4. | generation | n. | 产生，发电 |
| 5. | generate | v. | 产生，发生 |
| 6. | consumption | n. | 消费，消费量 |
| 7. | consume | vt. | 消费 |
| 8. | transmission | n. | 传输，输送，输电 |
| 9. | transmit | vt. | 传输，输送 |
| 10. | distribution | n. | 分配，分布，配电 |
| 11. | distribute | vt. | 分配，分布 |
| 12. | plant | n. | 电厂，工厂 |
| 13. | interconnection | n. | 互联 |
| 14. | load | n. | 负荷 |
| 15. | transmission line | n. | 输电线 |
| 16. | substation | n. | 分站，变电所 |
| 17. | switching | n. | 配电，变换，开关 |
| 18. | sub-transmission | n. | 中（等电）压输电 |
| 19. | source | n. | 电源，能源，源极 |
| 20. | redundancy | n. | 冗余 |
| 21. | contingency | n. | 偶然，可能性，意外事故 |
| 22. | service | n. | 供电；服务 |
| 23. | disruption | n. | 中断，瓦解，破坏 |
| 24. | topology | n. | 拓扑结构 |
| 25. | equipment | n. | 设备，装备，装置 |
| 26. | radial | adj. | 放射状的，半径的 |
|  |  | n. | 光线，射线 |
| 27. | loop | n. | 环，回路，回线 |
| 28. | power source |  | 电源 |
| 29. | section | n. | 截面，断面；部分 |
| 30. | circuit | n. | 电路 |
| 31. | feed | vt. | 供给，馈送，加载 |
| 32. | alternating | adj. | 交变的，交互的 |
| 33. | voltage | n. | 电压 |
| 34. | transform | v. | 改变，转化，变换 |
| 35. | flexibility | n. | 灵活性 |
| 36. | electric machine |  | 电机 |
| 37. | three-phase |  | 三相 |
| 38. | single-phase |  | 单相 |
| 39. | operate | v. | 操作，运转，运行 |
| 40. | balanced | adj. | 平衡的 |
| 41. | synchronous | adj. | 同步的，同时的 |
| 42. | synchronous machine |  | 同步电机，同步机 |
| 43. | subsystem | n. | 子系统 |
| 44. | convert | vt. | 使转变，转换 |
| 45. | transport | vt. | 传送，运输 |
| 46. | arc lamp | n. | 弧光灯 |
| 47. | lighthouse | n. | 灯塔 |

# Chapter 1　Basic Concepts of Electric Power System

| 48. | illumination | *n.* | 照明 |
| 49. | polyphase | *adj.* | 多相的 |
| 50. | reliability | *n.* | 可靠性 |

## PHRASES

**1. win out over: 战胜，胜过；在竞争或比较中取得优势，超过对手或其他事物**

Example: The AC system had won out over the DC system for the following reasons.
由于下列原因，交流系统已经完全战胜直流系统。

**2. be comprised of: 由…构成，包含…**

Example: Electric power systems are all comprised of three-phase AC systems operating essentially at constant voltage.
电力系统都是由基本运行在恒定电压的三相交流系统构成。

**3. so as to: 使得，以致**

Example: Single-phase residential and commercial loads are distributed equally among the phases so as to effectively form a balanced three-phase system.
单相民用负载和商业负载在各相间均匀分配，以有效地构成平衡的三相系统。

**4. play a role: 扮演…角色**

Example: The electric power system is one of the tools for converting and transporting energy which is playing an important role in meeting this challenge.
电力系统是在应对该挑战中扮演重要角色的转换和运输能量的工具之一。

**5. put into operation: 投入运行**

Example: The first AC transmission line in the United States was put into operation in 1890.
美国的第一条交流输电线于1890年投入运行。

**6. for instance: 相当于 for example**

Example: The operating voltage increased rapidly. For instance, in the United States, in 1890 the first line was operated at 3.3kV. Voltage rose to 100kV in 1907, 150kV in 1913, ..., and 287kV in 1936.
电压等级快速升高。例如，在美国，1890年首条线路运行在3.3千伏，1907年电压升高到100千伏，1913年为150千伏……1936年为287千伏。

**7. large blocks of: 大批量的**

Example: The demand for large blocks of power and increased reliability suggested the interconnection of neighboring systems.
对大批量电能的需求和可靠性的提高促进了相邻系统的互联。

**8. to the point of 或 to the point where: 达到…程度**

Example: Interconnection has increased to the point where power is exchanged between the systems of different companies as a matter of routine.

互联已经发展到这样的程度，功率在不同电力公司的系统之间交换已经成为惯例。

**9. as a matter of routine: 作为例行公事；按常规**

Example: 参见上例。

## COMPLICATED SENTENCES

**1. All the electricity the consumers take from the power system is generated in a power plant, or called power station, of some kind.**

译文：用户从电力系统取用的所有电能都是在某种形式的电厂（或者称为电站）中产生的。

说明：the consumers take from the power system 是 electricity 的定语从句，or called power station 是 power plant 的另一种说法，而 of some kind 也是直接修饰 power plant 的。为了便于理解，这个句子可以拆解为：

The consumers take electricity from the power system; all the electricity is generated in a power plant of some kind; the power plant is also called power station.

**2. Transmission systems are the connecting links between the generation systems and the distribution systems and lead to other power systems over interconnections.**

译文：输电系统连接发电和配电系统，并经过网络互联通向其他电力系统。

说明：这里 transmission system 是主语，are 和 lead 是并列谓语，通过第二个 and 连接。在翻译时为了语句通顺，名词动词化，把 connecting links 翻译成连接。为了便于理解，这个句子可以拆解为：

Transmission systems connect the generation systems and the distribution systems; transmission systems also lead to other power systems over interconnections.

**3. This provides a high degree of structural redundancy that enables the system to withstand unusual contingencies without service disruption to the consumers.**

译文：这提供了高度的结构化冗余，使系统能够承受罕见的意外事故而不会中断对用户的供电。

说明：从 that 开始到句子结束是 redundancy 的定语从句，在翻译时为了句子通顺，可以不作为定语处理。介词 without 表示在前面的情况发生时其后面的情况不同时发生，可以翻译为"而不会…"。service disruption to the consumers 可以直接译为"对用户的供电中断"，但为了语句通顺，最好处理为 to disrupt service to the consumers。为了便于理解，这个句子可以拆解为：

This provides a high degree of structural redundancy; the redundancy enables the system to withstand unusual contingencies; the system will not disrupt service to the consumers when

# Chapter 1　Basic Concepts of Electric Power System

withstanding contingencies.

**4. The equipment which forms an electric system is arranged depending on the manner in which load grows in the area and may be rearranged from time to time.**

译文：构成电力系统的设备，其布局取决于当地的负荷增长方式，并且随时可能重新调整。

说明：本句的主结构为 The equipment ... is arranged ... and may be rearranged ...。which forms an electric system 是 equipment 的定语，in which load grows in the area 是 manner 的定语。

**5. Development of sources of energy to accomplish useful work is the key to industrial progress which is essential to the continual improvement in the standard of living of people everywhere.**

译文：作有用功的能源的开发是工业发展的关键，而工业发展是人们生活水平不断提高所必需的。

说明：这是一个复合句，which is ... 解说的是 industrial progress。本句的主语是 development，is 做表语，宾语是 the key。Development of sources of energy to accomplish useful work 中，to accomplish useful work 修饰 energy, energy 修饰 sources, sources 修饰 development。

**6. The industry, by some standards, is the largest in the world.**

译文：按照某种标准，该行业是世界上最大的。

说明：by some standards 是插入语，意为"按照某种标准，依据特定的标准"。插入语在翻译时一般应提到句首。

**7. In 1953 came the first 345kV line.**

译文：1953年出现了第一条345kV线路。（1953年第一条345kV线路出现了。）

说明：这是一个倒装句，主语是 the first 345kV line。当主语较长而谓语又很短时，为了避免头重脚轻，句子往往用主谓倒装的形式来表达。也可以顺序写为 The first 345kV line came in 1953.

## ABBREVIATIONS (ABBR.)

| | | | |
|---|---|---|---|
| 1. | DC | direct current | 直流（电） |
| 2. | AC | alternating current | 交流（电） |

## SUMMARY OF GLOSSARY

| | | |
|---|---|---|
| 1. | (electric) power system | 电力系统 |
| | power generation | 发电 |
| | transmission system | 输电系统 |
| | sub-transmission system | 中（等电）压输电系统 |

|   |                       |            |
|---|-----------------------|------------|
|   | transmission network  | 输电网络   |
|   | distribution system   | 配电系统   |
| 2.| **power generation**  | **发电**   |
|   | power plant           | 电厂，发电厂 |
|   | power station         | 电厂，电站 |
| 3.| **load classification** | **负荷分类** |
|   | industrial loads      | 工业负荷   |
|   | residential loads     | 居民负荷   |
|   | commercial loads      | 商业负荷   |
| 4.| **system topology**   | **拓扑结构** |
|   | radial system         | 辐射状系统 |
|   | loop system           | 环状系统   |
|   | network system        | 网状系统   |

# EXERCISES

**1. Translate the following words or expressions into Chinese.**

(1) power generation　　(2) to the point of　　(3) transmission line

(4) power plant　　(5) load　　(6) single-phase

(7) service disruption　　(8) substation　　(9) power source

(10) for instance　　(11) as a manner of routine

**2. Translate the following words or expressions into English.**

(1) 电站　　(2) 变电站　　(3) 配电系统

(4) 互联　　(5) 功率交换　　(6) 环状系统

(7) 电能　　(8) 消耗　　(9) 设备

**3. Fill in the blanks with proper words or expressions.**

(1) All the electricity the consumers take from the power system is generated in a power plant, or called ＿＿＿＿＿, of some kind.

(2) Transmission systems connect the generation systems and the ＿＿＿＿＿ systems; they also lead to other power systems over ＿＿＿＿＿.

(3) This provides a high degree of structural redundancy that enables the system to withstand unusual contingencies without service ＿＿＿＿＿ to the consumers.

# Word-Building (1)　General Introduction
# 专业词汇和构词方法

在科学技术的各个领域，文献资料中都存在大量的专业词汇（仅用于某一学科或与

# Chapter 1　Basic Concepts of Electric Power System

其他学科通用的词汇或术语)。对专业词汇的快速认读和正确理解，是提高阅读速度和增强理解能力的关键，也是阅读和翻译相关领域专业技术文献的必要知识积累。

专业词汇的形成主要有三种情况：

(1) 借用日常英语词汇或其他学科的专业词汇，但是词义和词性可能发生了明显的变化。

例如：在日常英语中表示"力量，权力"和在机械专业表示"动力"的 power，在电力专业领域可以仍作为名词，表示"电力，功率，电能"；也可以作为动词，表示"供以电能"。在日常英语中表示"植物"的 plant 一词，在电力专业领域中用来表示"电厂"，等等。

(2) 由日常英语词汇或其他学科的专业词汇，直接合成新的词汇。

例如：over 和 head 组合成 overhead，表示"架空（输电线）"；super 和 conductor 组合成 superconductor，表示"超导体"。

| over | + | head | = | overhead | overhead lines |
|---|---|---|---|---|---|
| 在…上 |  | 头 |  | 在头上的，高架的 | 架空线 |
| super | + | conductor | = | superconductor |  |
| 超 |  | 导体 |  | 超导体 |  |

(3) 由基本词根 (etyma) 和前缀 (prefixes) 或后缀 (suffixes) 组成新的词汇。大部分专业词汇属于这种情况。一般来说，加前缀改变词意，加后缀改变词性。

例如：

| 词根 | develop | *vt.* | 发达，发展，开发 |
|---|---|---|---|
| 加后缀 -ed | developed | *adj.* | 发达的 |
| 再加前缀 un- | undeveloped | *adj.* | 不发达的，未开发的 |

了解专业英语的构词方法，有助于扩大词汇量，提高阅读能力。各章后面提供了一些比较基本的专业词汇构词方法和示例，供读者参考。

# Chapter 2
# Electric Power Generation

## Part 1  About Power Generation

Power generation is the first stage of electric energy application. Power systems use synchronous machines for the generation of electricity, in other words, synchronous generators form the principal source of electric energy in power systems. The synchronous machine as an AC generator driven by a turbine is the device which converts mechanical energy to electrical energy, with powers ranging up to 1500MW.* Prime movers convert the primary sources of energy to mechanical energy that is, in turn, converted to electrical energy by a synchronous generator.

Before connecting a generator to an infinite bus (or in parallel with another generator), it must be synchronized. A generator is said to be synchronized when it meets all the following conditions:

(1) The generator frequency is equal to the system frequency.
(2) The generator voltage is equal to the system voltage.
(3) The generator voltage is in phase with the system voltage.
(4) The phase sequence of the generator is the same as that of the system.

## Part 2  Hydropower Generation

When flowing water is captured and turned into electricity, it is called hydroelectric power or hydropower. Hydropower, also called water power, is a type of important common power source. Hydropower generating stations convert the energy of moving water into electrical energy by means of a hydraulic turbine coupled to a synchronous generator.

Hydropower offers advantages over other energy sources but faces unique environmental challenges.

(1) **Advantages:** Hydropower is fueled by water, so it's a clean fuel source. Hydropower doesn't pollute the air. Hydropower is a domestic source of energy, produced in many countries. Hydropower relies on the water cycle, which is driven by the sun, thus it's a renewable power source. Hydropower is generally available as needed; engineers can control the flow of water through the turbines to produce electricity on demand.

## Chapter 2  Electric Power Generation

Hydropower plants provide benefits <u>in addition to</u> clean electricity. These benefits may include a variety of recreational opportunities, water supply, and flood control.

(2) **Disadvantages:** Hydropower can impact water quality and flow. Hydropower plants can cause low dissolved oxygen levels in the water, a problem that is harmful to riverbank habitats.* Maintaining minimum flows of water downstream of a hydropower installation is also critical for the survival of riverbank habitats. Hydropower plants can be impacted by drought. When water is not available, the hydropower plants can't produce electricity. New hydropower facilities impact the local environment and may compete with other uses for the land. Those alternative uses may be more highly valued than electricity generation.

The hydraulic resources of most modern connotations are already fully developed. For instance, in the United States, water power <u>accounts for</u> less than 20% of the total and that percentage will drop because most of the available sources of water power have been developed.*

## Part 3  Thermal Generation

Thermal generating stations produce electricity from the heat released by the combustion of fossil fuels, such as coal, oil, or natural gas. Most stations have ratings between 200MW and 1500MW so as to attain the high efficiency and economy of a large installation.* Such a station has to be seen to appreciate its enormous complexity and size.

Thermal stations are usually located near a river or lake because large quantities of cooling water are needed to condense the steam as it exhausts from the turbines.*

The efficiency of thermal generating stations is always low because of the inherent low efficiency of the turbines.

Now most of the electric power is generated in steam-turbine plants. Coal is the most widely used fuel for steam plants. Many plants converted to oil in the early 1970s, but in the face of the continuous escalation in the price of oil and the necessity of reducing dependence on foreign oil, reconversion from oil to coal has taken place wherever possible. Gas turbines are used to a minor extent for short periods when a system is carrying a peak load.

## Part 4  Nuclear Generation

Nuclear stations produce electricity from the heat released by a nuclear reaction. When the nucleus of an atom splits in two (a process called atomic fission), a considerable amount of energy is released.

A nuclear station is similar to a conventional thermal station. The overall efficiency is also similar (between 30 and 40 percent), and a cooling system must be provided for it. Consequently, nuclear stations are also located close to rivers and lakes. In dry areas, cooling towers are installed.

Nuclear plants fueled by uranium account for a continually increasing share of the load, but their construction is slow and uncertain because of the difficulty of raising capital to meet the sharply rising cost of construction, constantly increasing safety requirements which cause redesign, public opposition to the operation of the plants, and delays in licensing.*

The supply of uranium is limited, but the fast breeder reactors now prohibited in the United States, have greatly extended the total energy available from uranium in Europe. Nuclear fusion is the great hope for the future. As a controllable fusion process on a commercial scale becomes feasible, it will continue to grow and take over the direct fuel applications. For instance, the electric car will probably be used in order to reserve fossil fuels (including petroleum and gas synthesized from coal) for aircraft and long-distance trucking.

Finally, in producing energy by any means, the protection of our environment is extremely important.

## Part 5  Application of Various Power Stations

The total power drawn by the customers of a large utility system fluctuates between wide limits, depending on the seasons and time of the day.

These power blocks give rise to three types of generating stations:

(1) Base-power stations that deliver full power at all times. Nuclear stations and coal-fired stations are particularly well adapted to furnish base demand.

(2) Intermediate-power stations that can respond relatively quickly to changes in demand, usually by adding or removing one or more generating units. Hydropower stations are well adapted for this purpose.

(3) Peak-generating stations that deliver power for brief intervals during the day. Such stations must be put into service very quickly. Consequently, they are equipped with prime movers such as diesel engines, gas turbines, or compressed-air motors that can be started up in a few minutes. In this regard, it is worth mentioning that thermal generating stations using gas or coal take from 4 to 8 hours to start up, while nuclear stations may take several days. Obviously, such generating stations cannot be used to supply short-term peak power.

Still, many other energies are used which contribute to the industries and daily life, which will be introduced in Chapter 24.

### NEW WORDS AND EXPRESSIONS

| | | | |
|---|---|---|---|
| 1. | stage | n. | 发展过程中的阶段，时期，步骤 |
| 2. | synchronous | adj. | 同步的，同速率运动或操作的 |
| 3. | principal | adj. | 主要的，首要的 |
| 4. | turbine | n. | （涡）轮机，叶轮机，汽轮机 |
| 5. | prime mover | | 原动机，原动力，牵引车，推进器 |

# Chapter 2　Electric Power Generation

| 6. | bus | n. | 母线，总线，汇流条 |
|---|---|---|---|
| 7. | synchronize | v. | 同步，使同步 |
| 8. | frequency | n. | 频率，周率，发生次数 |
| 9. | in phase | | 同相地，协调地 |
| 10. | phase sequence | | 相位序列，相序 |
| 11. | hydropower | n. | 水电，水力发出的电力 |
| 12. | hydroelectric | adj. | 水电的，水力发电的，水力电气的 |
| 13. | hydraulic | adj. | 水力的，水压的 |
| 14. | couple | v. | 连接，结合，耦合 |
| 15. | fuel | v. | 刺激，推动；加燃料，供燃料；得到燃料 |
| | | n. | 燃料，养料，动力 |
| 16. | recreational | adj. | 休养的，娱乐的 |
| 17. | hydropower plants | | 水电厂，水电站 |
| 18. | dissolved | adj. | 溶解的，溶化的，解散的 |
| 19. | oxygen | n. | 氧 |
| 20. | riverbank | n. | 河堤，河岸 |
| 21. | downstream | adj. | 下游的；adv. 下游地 |
| 22. | installation | n. | （整套）装置[备]，设备[施]；台，站；安装，装置 |
| 23. | connotation | n. | 含蓄，储蓄的东西 |
| 24. | thermal | adj. | 热的，热量的 |
| 25. | combustion | n. | 燃烧 |
| 26. | fossil | n. | 化石，僵化的事物 |
| | | adj. | 化石的，陈腐的，守旧的 |
| 27. | rating | n. | 额定，等级级别（尤指军阶） |
| 28. | condense | v. | （使）（气体）凝结，冷凝 |
| 29. | steam-turbine | | （蒸）汽轮机 |
| 30. | escalation | n. | 扩大；不断增加；逐步上升 |
| 31. | reconversion | n. | 恢复原状；再转变 |
| 32. | gas turbine | | 燃气轮机 |
| 33. | minor extent | | 小幅度，小尺寸 |
| 34. | peak load | | 最大负荷，峰荷 |
| 35. | nuclear station | | 核电站 |
| 36. | reaction | n. | 反应，反作用，感应 |
| 37. | fission | n. | 裂开，分体，裂变 |
| | | v. | （使）裂变 |
| 38. | cooling tower | | 冷却塔 |
| 39. | uranium | n. | 铀 |
| 40. | fast breeder reactor | | 快中子增殖反应堆 |
| 41. | nuclear fusion | | 核聚变 |
| 42. | feasible | adj. | 可行的，切实可行的 |
| 43. | petroleum | n. | 石油 |
| 44. | synthesize | v. | 综合，合成 |
| 45. | drawn | v. | draw 的过去分词，领取，提取 |

| 46. | utility | n. | 公用事业公司：如电力，水和交通 |
| 47. | base-power | | 基本功率；基荷 |
| 48. | coal-fired station | | 燃煤电厂 |
| 49. | furnish | v. | 供应；提供；供给 |
| 50. | intermediate | adj. | 中间的 |
| 51. | generating unit | | 发电机组 |
| 52. | diesel engine | | 柴油机 |
| 53. | compressed air | | 压缩空气 |

## PHRASES

**1. in other words: 换言之，换句话说**

Example: Power systems use synchronous machines for the generation of electricity, <u>in other words</u>, synchronous generators form the principal source of electric energy in power systems.

电力系统用同步电机产生电能，换句话说，同步发电机形成电力系统中电能的主要来源。

**2. in turn: 转而、随后**

Example: Prime movers convert the primary sources of energy to mechanical energy that is, <u>in turn</u>, converted to electrical energy by a synchronous generator.

原动机将一次能源转化为机械能，随后由发电机转化为电能。

**3. in parallel with: 与…并联，与…平行，与…同时**

Example: Before connecting a generator to an infinite bus (or <u>in parallel with</u> another generator), it must be synchronized.

在将一台发电机接入无穷大母线或与另一台发电机并联之前，必须使其同步。

**4. by means of: 依靠**

Example: Hydropower generating stations convert the energy of moving water into electrical energy <u>by means of</u> a hydraulic turbine coupled to a synchronous generator.

依靠与同步发电机相连接的水轮机，水电站将移动的水的能量转换为电能。

**5. in addition to: 除了…之外，还**

Example: Hydropower plants provide benefits <u>in addition to</u> clean electricity.

除了提供清洁电能之外，水电厂还提供其他好处。

**6. account for: 占（…比重），解决，说明，得分**

Example: In the United States, water power <u>accounts for</u> less than 20% of the total.

在美国，水电在全部电力中所占的比例不到20%。

**7. put into service: 交付使用，投入运行，使工作**

Example: Such stations must be <u>put into service</u> very quickly.

这种场站必须非常快地投入运行。

**8. In this regard: 在这点上**

Example: In this regard, it is worth mentioning that thermal generating stations using gas or coal take from 4 to 8 hours to start up, while nuclear stations may take several days.

在这点上，值得注意的是，采用天然气或煤炭的热电厂启动需要 4～8h，而核电站会花费几天时间。

## COMPLICATED SENTENCES

**1. The synchronous machine as an AC generator driven by a turbine is the device which converts mechanical energy to electrical energy, with powers ranging up to 1500MW.**

译文：作为由涡轮机驱动的交流发电机，同步电机是将机械能转化为电能的设备，其功率范围高达 1500MW。

说明：句子的主语中心词是 the synchronous machine，以 driven by a turbine 作为定语的 an AC generator 用来说明 the synchronous machine 的性质。Which 引导的定语从句用来说明 device 的功能。为了便于理解，这个句子可以拆解为：

The synchronous machine is an AC generator; the AC generator is driven by a turbine; the synchronous machine is a device; the device converts mechanical energy to electrical energy; the energy conversion is in powers ranging up to 1500MW.

**2. Hydropower plants can cause low dissolved oxygen levels in the water, a problem that is harmful to riverbank habitats.**

译文：水电厂会导致水中的氧气溶解度降低——（这是）一个对岸边生存环境有害的问题。

说明：作为 cause（引起，导致，造成）的宾语，low dissolved oxygen levels in the water 若直接翻译为"水中低的溶解氧的水平"显然不太合适，这种情况在翻译时可以将名词动词化，处理为"水中的氧气溶解度降低"就明显通顺了。a problem 与 low dissolved oxygen levels in the water 可以看作是同位语，即二者删掉一个不影响句子结构的完整性，翻译时可以用破折号表示二者的关联。为了便于理解，这个句子可以拆解为：

Hydropower plants can cause low dissolved oxygen levels in the water; low dissolved oxygen levels in the water is a problem that is harmful to riverbank habitats.

**3. For instance, in the United States, water power accounts for less than 20% of the total and that percentage will drop because most of the available sources of water power have been developed.**

译文：例如，在美国，水电在全部电力中所占的比例不到 20%，并且，因为大多数可用的水电资源已被开发，该百分比还会下降。

说明：because 引导的从句只是 that percentage will drop 的原因，不是 Water power accounts for less than 20% of the total 的原因，在翻译时要格外注意。有的人习惯翻译为"……水电在全部电力中所占的比例不到20%，并且该百分比还会下降，因为大多数可用的水电资源已被开发。"对于这个特定的例子，这样翻译当然读者也可能看懂，但是这种翻译风格容易发生歧义或混淆。为了便于理解，这个句子可以拆解为：

For instance, in the United States, water power accounts for less than 20% of the total; and because most of the available sources of water power have been developed that percentage will drop.

**4. Most stations have ratings between 200MW and 1500MW so as to attain the high efficiency and economy of a large installation.**

译文：大多数电站的额定值在 200 ～ 1500MW，以实现大站的高效经济（运行）。

说明：这个句子结构并不复杂，难点是如何流畅通顺地翻译为中文。为了更加符合中文的语法和书写习惯，建议采用意译的方式。

**5. Thermal stations are usually located near a river or lake because large quantities of cooling water are needed to condense the steam as it exhausts from the turbines.**

译文：热电厂通常位于河流或湖泊附近，这是因为需要大量的冷却水在蒸汽从汽轮机排出时将其冷凝（或者：……需要大量的冷却水去冷凝从汽轮机排出的蒸汽）。

说明：这是一个复合句，既有构成因果关系的 because 从句，又有作为条件状语的 as 从句，需要理清相互关系。这个句子阅读起来不算太难，关键是如何正确地用中文表达出来。

**6. Nuclear plants fueled by uranium account for a continually increasing share of the load, but their construction is slow and uncertain because of the difficulty of raising capital to meet the sharply rising cost of construction, constantly increasing safety requirements which cause redesign, public opposition to the operation of the plants, and delays in licensing.**

译文：以铀为燃料的核电厂，负责应对持续增长的负荷份额，但是其建设缓慢且不确定，原因包括：难以增加资金满足急剧增长的建设成本要求，不断增加的安全要求会导致重新设计，公众对电站运行的反对，以及经营许可的迟缓。

说明：句子结构不是很复杂，但是很长。在 because of 后面列举了多个原因，这些原因是并列关系。出现在 because of 后面的短语，在翻译时往往要做适当的语序调整，尤其是多个原因并列出现时，要尽量保持翻译风格的一致和语句的连贯通顺。

## SUMMARY OF GLOSSARY

1. **hydropower**      水电
   - hydropower      水力发出的电力
   - hydroelectric power      水力发电
   - water power      水电

# Chapter 2  Electric Power Generation

2. **fossil fuel** 化石燃料
   coal 煤
   oil, petroleum 石油
   gas, natural gas 天然气
3. **thermal plants** 热电厂
   fossil fuel plant 化石燃料电厂
   coal-fired plant 燃煤电厂
   steam-turbine plant 蒸汽轮机电厂
   gas-turbine plant 燃气轮机电厂
4. **nuclear reaction** 核反应
   nuclear fission 核裂变
   nuclear fusion 核聚变
   fast breeder reactor 快速中子反应堆

# EXERCISES

**1. Translate the following words or expressions into Chinese.**

(1) synchronous        (2) prime mover        (3) turbine
(4) infinite bus       (5) efficiency         (6) phase sequence
(7) hydroelectric      (8) fossil fuel        (9) installation
(10) facilities        (11) nuclear fusion    (12) generating unit

**2. Translate the following words or expressions into English.**

(1) 发电机              (2) 峰荷                (3) 蒸汽轮机
(4) 同相                (5) 同步                (6) 燃烧
(7) 核裂变              (8) 投入运行            (9) 公用事业公司

**3. Fill in the blanks with proper words or expressions.**

(1) The synchronous generator driven by a turbine is the device which converts _____ energy to _____ energy, with powers ranging up to 1500MW.

(2) Thermal stations are usually located near a river or lake because large quantities of _____ are needed to condense the steam as it exhausts from the _____.

(3) Hydropower plants provide other benefits _____ clean electricity.

# Word-Building (2)  electro-, magneto  电，磁

**1. electr-，electri-，electro-  表示：电，电的**

electric        *adj.*   电的，用电的，电动的，发电的
electrical      *adj.*   电的，有关电的

| electrician | n. | 电工，电气技师 |
| electrion | n. | 高压放电 |
| electronic | adj. | 电子的 |
| electromagnetic | adj. | 电磁的 |
| electromechanical | adj. | 机电的，电动机械的 |

## 2. magnet-，magneto-　表示：磁，磁性，磁力

| magnetics | n. | 磁学，磁性元件，磁性材料 |
| magnetize | v. | 使磁化，励磁，受磁 |
| magnetization | n. | 磁化，磁化强度 |
| magnetic(al) | adj. | 磁的，有磁性的，有吸引力的 |
| magnetoresistor | n. | 磁控电阻器 |
| magnetoconductivity | n. | 导磁性 |
| magnetoelectric | adj. | 磁电的 |

# Chapter 3
# Power Generating Stations

## Part 1  Hydropower Stations

Hydropower generating stations convert the energy of moving water into electrical energy by means of a hydraulic turbine coupled to a synchronous generator. They are all powered by the kinetic energy of flowing water as it moves downstream.

There are three types of hydropower facilities. The most common type of hydroelectric power plant is an *impoundment facility*, typically a large hydropower system that uses a dam to store river water in a reservoir. Water released from the reservoir flows through a turbine, spinning it, which in turn activates a generator to produce electricity.* The water may be released either to meet changing electricity needs or to maintain a constant reservoir level.

Such a hydropower installation consists of dams, waterways, and conduits that form a reservoir and channel the water toward the turbines.* These, and other items described, enable us to understand some of the basic features and components of a hydropower station, as shown in Figure 3.1.

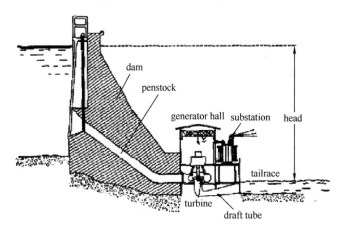

**Figure 3.1**  Cross-section view of a medium-head hydropower station

Dams made of earth or concrete are built across river beds to create storage reservoirs. Reservoirs can compensate for the reduced precipitation during dry seasons and for the abnormal flows that accompany heavy rains and melting snow.* Dams permit us to regulate the water flow throughout the year, so that the powerhouse may run at <u>close to</u> full capacity.

Spillways <u>adjacent to</u> the dam are provided to discharge water whenever the reservoir level is too high.

In large installations, conduits lead the water from the dam site to the generating plant. The penstocks channel the water into a scroll-case. Guide vanes and wicket gates control the water so that it flows smoothly into the runner blades. The wicket gates open and close <u>in response to</u> a powerful hydraulic mechanism.

Water that has passed through the runner blades moves next through a carefully designed vertical channel, called a draft tube. The draft tube improves the hydraulic efficiency of the turbine. It leads out to the tailrace, which channels the water into the downstream river bed.

The power house contains the synchronous generators and associated control apparatus, etc. Finally, still, many other devices (too numerous to mention here) make up the complete hydropower station.

A *diversion*, sometimes called *run-of-river*, *facility* channels a portion of a river through a canal or penstock. It may not require the use of a dam. Figure 3.2 is a photo of an aerial view of a river with a waterfall and no dam. The hydropower intake and outlet are labeled. The intake is above the waterfall, the outlet is below it.

**Figure 3.2**　The Tazimina project in Alaska

When the demand for electricity is low, a *pumped storage facility* stores energy by pumping water from a lower reservoir to an upper reservoir. During periods of high electrical demand, the water is released back to the lower reservoir to generate electricity.

Hydropower stations are divided into three groups depending on the head of water. The "head" refers to the difference of elevation between the upper reservoir above the turbine and the tailrace of the discharge point just below the turbine.* *High-head developments* have heads in excess of 300m, and high-speed turbines are used. The amount of impounded water is usually small. *Medium-head developments* have heads between 30m and 300m, and medium-speed Francis turbines are used. The generating station is fed by a huge reservoir of water retained by dikes and a dam. *Low-head developments* have heads under 30m, and low-speed turbines are used. These generating stations often extract energy from flowing rivers. The turbine are designed to handle large volumes of water at low pressure. No reservoir is provided.

# Chapter 3  Power Generating Stations

Hydropower plants range in size from small systems for a home or village to large projects producing electricity for utilities. *Large hydropower*, *small hydropower* and, *micro hydropower* are defined for different sizes of hydropower plants.

## Part 2  Thermal Stations

Now thermal stations, called heat-engine plants, are the most common power plants in the world. In a heat-engine plant, fossil fuels, such as coal, oil, or natural gas, are used as primary sources.

The basic structure and principal components of a thermal generating station are shown in Figure 3.3. They are itemized and described below.

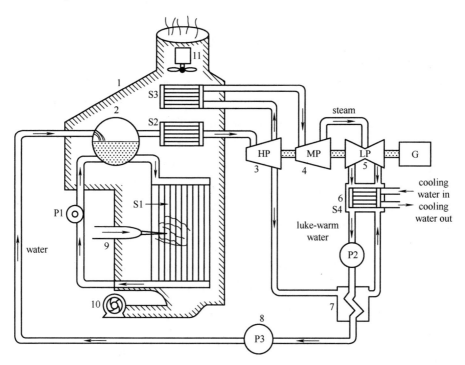

**Figure 3.3**  Principal components of a thermal power plant

• A huge boiler (1) acts as a furnace, transferring heat from the burning fuel to row upon row of water tubes S1, which entirely surround the flames.* Water is kept circulating through the tubes by a pump P1.

• A drum (2) containing water and steam under high pressure produces the steam required by the turbines. It also receives the water delivered by boiler-feed pump P3. Steam races toward the high-pressure turbine HP after having passed through superheater S2. The superheater, composed of a series of tubes surrounding the flames, raises the steam temperature ensures that the steam is absolutely dry, and raises the overall efficiency of the station.

• A high-pressure (HP) turbine (3) converts thermal energy into mechanical energy by letting the steam expand as it moves through the turbine blades. The temperature and pressure at the output of the turbine are, therefore, less than at the input. In order to raise the thermal efficiency and prevent premature condensation, the steam passes through a reheater S3, composed of a third set of heated tubes.

- The medium-pressure (MP) turbine (4) is similar to the high-pressure turbine, except that it is bigger so that the steam may expand still more.
- The low-pressure (LP) turbine (5) is composed of two identical left-hand and right-hand sections. The turbine sections remove the remaining available energy from the steam. The steam flowing out of LP expands into an almost perfect vacuum created by the condenser (6).
- Condenser (6) causes the steam to condense by letting it flow over cooling pipes S4. Cold water from an outside source, such as a river or lake, flows through the pipes, thus carrying away the heat. It is the condensing steam that creates the vacuum.

A condensate pump P2 removes the lukewarm condensed steam and drives it through a reheater (7) toward a feedwater pump (8).

- The reheater (7) is a heat exchanger. It receives hot steam, bled off from a high-pressure turbine HP, to raise the temperature of the feedwater. Thermodynamic studies show that the overall thermal efficiency is improved when some steam is bled off this way, rather than letting it follow its normal course through all three turbines.
- The burners (9) supply and control the amount of gas, oil, or coal injected into the boiler. Coal is pulverized before it is injected. Similarly, heavy bunker oil is preheated and injected as an atomized jet to improve surface contact (and combustion) with the surrounding air.
- A forced-draft fan (10) furnishes the enormous quantities of air needed for combustion.
- An induced-draft fan (11) carries the gases and other products of combustion toward the cleansing apparatus and from there to the stack and the outside air.
- Generator G, directly coupled to all three turbines, converts the mechanical energy into electrical energy.

In practice, a steam station has hundreds of components and accessories to ensure high efficiency, safety, and economy. For example, control valves regulate the amount of steam flowing to the turbines; complex water purifiers maintain the required cleanliness and chemical composition of the feedwater; oil pumps keep the bearings properly lubricated. However, the basic components we have just described enable us to understand the operation and some of the basic problems of a thermal station.

# Part 3   Nuclear Stations

Nuclear stations produce electricity from the heat released by a nuclear reaction. When the nucleus of an atom splits in two (a process called atomic fission), a considerable amount

# Chapter 3   Power Generating Stations

of energy is released.

A nuclear station is identical to a thermal station except that the boiler is replaced by a nuclear reactor. The reactor contains the fissile material that generates the heat. A nuclear station, therefore, contains a synchronous generator, steam turbine, condenser, and so on, similar to those found in a conventional thermal station. The overall efficiency is also similar (between 30 and 40 percent), and a cooling system must be provided for it. Consequently, nuclear stations are also located close to rivers and lakes. In dry areas, cooling towers are installed. Owing to these similarities, we will only examine the operating principle of the reactor itself.

In the case of a nuclear reactor, we have to slow down the neutrons to increase their chance of striking other uranium nuclei. Toward this end, small fissionable masses of uranium fuel are immersed in a moderator. The moderator may be ordinary water, heavy water, graphite, or any other material that can slow down neutrons without absorbing them. By using an appropriate geometrical distribution of the uranium fuel within the moderator, the speed of the neutrons can be reduced so they have the required velocity to initiate other fusions.* Only then will a chain reaction take place, causing the reactor to go critical.*

As soon as the chain reaction starts, the temperature rises rapidly. To keep it at an acceptable level, a liquid or gas has to flow rapidly through the machine to carry away the heat.* This coolant may be heavy water, ordinary water, liquid sodium, or a gas like helium or carbon dioxide. The hot coolant moves in a closed circuit which includes a heat exchanger. The latter transfers the heat to a steam generator that drives the turbines, as shown in Figure 3.4. Thus, contrary to what its name would lead us to believe, the coolant is not cool but searingly hot.*

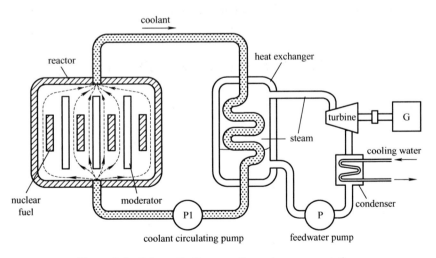

**Figure 3.4**   Schematic diagram of a nuclear power station

There are several types of reactors, but the following are the most important:

(1) Pressure-Water Reactor (PWR). Water is used as a coolant and it is kept under such high pressure that it cannot boil off into steam. Ordinary water, as in light-water reactors may be used, or heavy water, as in Canada's Deuterium Uranium reactors.

(2) Boiling-Water Reactors (BWR). The coolant in this reactor is ordinary water boiling under high pressure and releasing steam. This eliminates the need for a heat exchanger because the steam circulates directly through the turbines. However, as in light-water reactors, enriched uranium dioxide must be used containing about 3 percent $U^{235}$.

(3) High-Temperature Gas Reactor (HTGR). This reactor uses an inert gas coolant such as helium or carbon dioxide. Due to the high operating temperature (typically 750℃ ), graphite is used as a moderator. The steam created by the heat exchanger is as hot as that produced in a conventional coal-fired steam boiler. Consequently, the overall efficiency of HTGR stations is about 40 percent.

(4) Fast Breeder Reactor (FBR). This reactor has the remarkable ability to both generate heat and create additional nuclear fact while it is in operation.

We have seen that splitting the nucleus, or called *nuclear fission*, of a heavy element such as uranium results in a decrease in mass and a release of energy.* We can also produce energy by combining the nuclei of two light elements in a process called *nuclear fusion*. For example, energy is released by the fusion of an atom of deuterium with an atom of tritium. If both the atomic concentration and speed are high enough, a self-sustaining chain reaction will result.

Unfortunately, we run into almost insurmountable problems when we try to control the fusion reaction, as we must do in a nuclear reactor. Basically, scientists have not yet succeeded in confining and controlling high-speed particles without at the same time slowing them down.

A major worldwide research effort is being devoted to solving this problem. If scientists succeed in domesticating nuclear fusion, it could mean the end of the energy shortage because hydrogen is the most common element on Earth.

## NEW WORDS AND EXPRESSIONS

| | | | |
|---|---|---|---|
| 1. | hydraulic | adj. | 水力的，水压的 |
| 2. | kinetic energy | | 动能 |
| 3. | impoundment | n. | 蓄水，积水 |
| 4. | reservoir | n. | 水库，蓄水池 |
| 5. | spin | vt. | 使快速旋转；扭转 |
| 6. | conduit | n. | 管道，导管；沟渠；泉水，喷泉 |
| 7. | concrete | n. | 混凝土 |
| 8. | precipitation | n. | 降水，降水量；沉淀 |
| 9. | powerhouse | n. | 发电站 |
| 10. | spillway | n. | 溢洪道，泄洪道 |
| 11. | discharge | n.v. | 卸下；放出；排水；放电 |
| 12. | penstock | n. | 水道，水渠，压力水管，水阀门 |
| 13. | scroll-case | n. | 漩涡形箱体 |
| 14. | guide vane | | 导（流叶）片，导向叶片 |
| 15. | wicket gate | | 水闸门，导水门，导叶 |

# Chapter 3　Power Generating Stations

| 16. | runner blade | | 主动轮叶 |
|---|---|---|---|
| 17. | draft tube | | 通风管，引流管，尾水管 |
| 18. | tailrace | n. | （水轮，涡轮的）放水路 |
| 19. | apparatus | n. | 器械，设备，仪器 |
| 20. | head of water | | 水头，落差 |
| 21. | elevation | n. | 海拔；仰角；提高；正面图 |
| 22. | in excess of | | 超过，较…为多 |
| 23. | impound | vt. | 蓄水；储存（水等）；灌溉 |
| 24. | Francis turbine | | 轴同辐流式涡轮机 |
| 25. | dike | n. | 堤防 |
| 26. | diversion | n. | 导流，分出，引出 |
| 27. | waterfall | n. | 瀑布，瀑布似的东西 |
| 28. | intake | n. | 入口，进口，通风口 |
| 29. | outlet | n. | 出口，出路 |
| 30. | pumped storage | | 抽水蓄能 |
| 31. | heat-engine | n. | 热力发动机 |
| 32. | primary source | | 原始来源，一次能源 |
| 33. | boiler | n. | 汽锅，锅炉 |
| 34. | furnace | n. | 炉子，熔炉 |
| 35. | flame | n. | 火焰 |
| 36. | pump | n. | 泵，抽水机 |
| 37. | drum | n. | 鼓，鼓形圆桶 |
| 38. | race | vi. | 迅速行进 |
| 39. | superheater | n. | 过热设备，过热器 |
| 40. | premature | adj. | 过早的 |
| 41. | reheater | n. | 再热器，回热器 |
| 42. | vacuum | n. | 真空，空间 |
| 43. | condenser | n. | 冷凝器，电容器 |
| 44. | lukewarm | adj. | 微温的 |
| 45. | feedwater | n. | 给水 |
| 46. | thermodynamic | adj. | 热力学的，使用热动力的 |
| 47. | burner | n. | 火炉，烧火的人 |
| 48. | pulverize | v. | 研磨成粉 |
| 49. | bunker | n. | 燃料，燃料仓 |
| 50. | preheat | vt. | 预先加热 |
| 51. | atomized jet | | 喷雾，雾状的喷射 |
| 52. | combustion | n. | 燃烧 |
| 53. | forced draft | | 强迫通风 |
| 54. | induced draft | | 进气通风，诱导通风 |
| 55. | cleansing apparatus | | 去污设备 |
| 56. | stack | n. | 烟囱或烟道 |
| 57. | purifier | n. | 清洁器 |
| 58. | bearing | n. | 轴承 |
| 59. | lubricate | vt. | 给…涂油；使润滑 |
| 60. | nucleus | n. | 核子；复数为 nuclei |

| 61. | fission | n.&v. | （使）裂变 |
| 62. | reactor | n. | 反应堆；电抗器 |
| 63. | fissile | adj. | 分裂性的，易分裂的，裂变的 |
| 64. | fissionable | n. | 可裂变物质 |
|  |  | adj. | 可引起核分裂的，可分裂的 |
| 65. | mass | n. | 团，块；没有具体形状的物质 |
| 66. | moderator | n. | （原子反应堆的）减速剂，慢化剂 |
| 67. | graphite | n. | 石墨 |
| 68. | fusion | n. | 合成；聚变，核合成 |
| 69. | chain reaction |  | 连锁反应 |
| 70. | coolant | n. | 冷冻剂，冷却液，散热剂 |
| 71. | liquid sodium |  | 液体钠 |
| 72. | helium | n. | 氦（化学元素，符号为 He） |
| 73. | carbon dioxide |  | 二氧化碳 |
| 74. | searingly | adv. | 灼热，烫人地 |
| 75. | schematic diagram |  | 原理图，示意图 |
| 76. | boil off |  | 蒸发损耗 |
| 77. | eliminate | v. | 排除，消除 |
| 78. | enriched uranium dioxide |  | 浓缩二氧化铀 |
| 79. | inert gas |  | 惰性气体 |
| 80. | deuterium | n. | [化] 氘 |
| 81. | tritium | n. | [化] 氚（氢的放射性同位素） |
| 82. | concentration | n. | 浓缩，浓度；集中，集合，专心 |
| 83. | self-sustaining | adj. | 自给的，自立的；自保持的 |
| 84. | insurmountable | adj. | 不能克服的，不能超越的 |
| 85. | particle | n. | 粒子，微粒，质点 |
| 86. | domesticate | vt. | 驯化；使归化；使通俗化； |

# PHRASES

**1. close to:** （数量）接近，（位置）靠近

Example: Dams permit us to regulate the water flow throughout the year, so that the powerhouse may run at close to full capacity.

水坝使我们可以调节全年的水流，从而使发电厂可以接近满负荷运行。

**2. adjacent to:** （位置）临近，靠近

Example: Spillways adjacent to the dam are provided to discharge water whenever the reservoir level is too high.

临近水坝的泄洪道用来在水库水位太高时放水。

**3. in response to:** 响应，适应；根据…行动，由…控制

Example: The wicket gates open and close in response to a powerful hydraulic mechanism.

水闸门的打开和关闭由一个强有力的液压装置控制。

# Chapter 3　Power Generating Stations

## COMPLICATED SENTENCES

**1. Water released from the reservoir flows through a turbine, spinning it, which in turn activates a generator to produce electricity.**

译文：从水库释放的水流过涡轮机，使其高速旋转，随后带动发电机产生电能。

说明：句中 it 代指 turbine。which in turn activates a generator to produce electricity 也是 turbine 的定语。在翻译时可以按照句子顺序翻译，注意各部分的相互关系就可以了。为了便于理解，这个句子可以拆解为：

Water released from the reservoir flows through a turbine; water spins the turbine; the turbine in turn activates a generator to produce electricity.

**2. Such a hydropower installation consists of dams, waterways, and conduits that form a reservoir and channel the water toward the turbines.**

译文：这种水力发电装置由水坝、水路和引水渠组成，这些形成水库并将水引至涡轮机。

说明：词组 consist of 表示"由…构成，包括…"，that 引导的定语从句修饰的主体是 dams, waterways, and conduits。为了便于理解，这个句子可以拆解为：

Such a hydropower installation consists of dams, waterways, and conduits; dams, waterways, and conduits form a reservoir and channel the water toward the turbines.

**3. Reservoirs can compensate for the reduced precipitation during dry seasons and for the abnormal flows that accompany heavy rains and melting snow.**

译文：水库可以补偿旱季减少的降水量以及伴随暴雨和融雪发生的异常水流。

说明：during dry seasons 是 the reduced precipitation 的定语，that 引导的定语从句修饰的主体是 the abnormal flows。句子谓语 compensate for 带的宾语有两个，一是 the reduced precipitation，一是 the abnormal flows。为了便于理解，这个句子可以拆解为：

Reservoirs can compensate for the reduced precipitation during dry seasons; and reservoirs can compensate for the abnormal flows; the abnormal flows accompany heavy rains and melting snow.

**4. The "head" refers to the difference of elevation between the upper reservoir above the turbine and the tailrace of the discharge point just below the turbine.**

译文："水头"指的是轮机上方的上游水库与轮机下方的排水地点之间的海拔高度的差别（落差）。

说明：这个句子结构并不复杂，但是涉及很多专业词汇，翻译时要注意。

**5. A huge boiler (1) acts as a furnace, transferring heat from the burning fuel to row upon row of water tubes S1, which entirely surround the flames.**

译文：一个巨大的汽锅（1）作为火炉，将燃料燃烧的热量传递给完全被火焰包围的成排的水管 S1。

说明：有两个现成的词组，一是 row up，表示"使成排"；一是 row of tubes，表示"管排，成排的管子"。在这里 row upon row of 应该理解为"一排排的，成排的"，注意不要混淆。由 which 引导的定语从句修饰 S1。

**6. By using an appropriate geometrical distribution of the uranium fuel within the moderator, the speed of the neutrons can be reduced so they have the required velocity to initiate other fusions.**

译文：通过铀燃料在缓和剂中采用适当的几何分布，中子的速度可以降低，从而使其具有启动其他合成所需的速率。

说明：逗号前面的内容如果直译很拗口，需要按照中文的表述习惯对语序作适当的调整。

**7. Only then will a chain reaction take place, causing the reactor to go critical.**

译文：只有这样，才会发生链式反应，使反应堆变成临界[极限]状态。

说明：这是一个倒装句。

**8. To keep it at an acceptable level, a liquid or gas has to flow rapidly through the machine to carry away the heat.**

译文：为使其保持在可以接受的水平，必须有液体或气体迅速流过电机将热量带走。

**9. Thus, contrary to what its name would lead us to believe, the coolant is not cool but searingly hot.**

译文：因而，与其名称要使我们相信的内容恰恰相反，冷却剂不是凉的而是相当灼热的。

说明：would 是虚拟语态用词，并不代表过去时态。

**10. We have seen that splitting the nucleus, or called *nuclear fission*, of a heavy element such as uranium results in a decrease in mass and a release of energy.**

译文：我们已经看到，像铀这样的重元素的核子分裂——或称为核裂变——会导致质量减少并释放能量。

说明：a decrease in mass 和 a release of energy 在翻译时可以名词动词化。插入到句子中间的 or called *nuclear fission* 描述的内容与 splitting the nucleus 相同，译成中文时建议用破折号表示。

## ABBREVIATIONS (ABBR.)

| | | | |
|---|---|---|---|
| 1. | PWR | pressure-water reactor | 高压水反应堆，压水堆 |
| 2. | BWR | boiling-water reactors | 沸水反应堆 |
| 3. | HTGR | high-temperature gas reactor | 高温燃气反应堆 |
| 4. | FBR | fast breeder reactor | 快速中子反应堆 |

# Chapter 3　Power Generating Stations

## SUMMARY OF GLOSSARY

1. **types of hydropower plants**　　　　　　水电站类型
   impoundment facility　　　　　　　　　　蓄水电站
   diversion facility　　　　　　　　　　　　导流电站，分流电站
   run-of-river facility　　　　　　　　　　　导流电站（同上）
   pumped storage facility　　　　　　　　　抽水蓄能电站
2. **groups depending on the head of water**　根据水头分组
   high-head development　　　　　　　　　高水头（落差）开发
   medium-head development　　　　　　　中水头（落差）开发
   low-head development　　　　　　　　　低水头（落差）开发
3. **scales of hydropower**　　　　　　　　　水电的规模
   large hydropower　　　　　　　　　　　　大水电
   small hydropower　　　　　　　　　　　　小水电
   micro hydropower　　　　　　　　　　　　微型水电
4. **nuclear reaction**　　　　　　　　　　　核反应
   pressure-water reactor　　　　　　　　　　高压水反应堆
   boiling-water reactor　　　　　　　　　　沸水反应堆
   high-temperature gas reactor　　　　　　　高温燃气反应堆
   fast breeder reactor　　　　　　　　　　　快速中子反应堆

# EXERCISES

**1. Translate the following words or expressions into Chinese.**

(1) hydraulic　　　　　(2) powerhouse　　　　(3) apparatus
(4) fission　　　　　　(5) capacity　　　　　　(6) chain reaction
(7) the latter　　　　　(8) boiler　　　　　　　(9) self-sustaining

**2. Translate the following words or expressions into English.**

(1) 抽水蓄能　　　　　(2) 反应堆　　　　　　(3) 反常的
(4) 聚变　　　　　　　(5) 预先加热　　　　　(6) 燃烧
(7) 冷凝器　　　　　　(8) 短缺　　　　　　　(9) 回热器

**3. Fill in the blanks with proper words or expressions.**

(1) _____ generating stations convert the energy of moving water into electrical energy by means of a _____ turbine coupled to a synchronous _____.

(2) Now _____ stations, called _____ plants, are the most common power plants in the world. In such a plant, fossil fuels, such as coal, oil, or natural gas, are used as _____.

(3) A nuclear station is identical to a thermal station except that the _____ is replaced by a _____.

· 29 ·

# Word-Building (3) hydro-，thermo- 水，热

## 1. hydr，hydro- 表示：水，液，流体

| | | |
|---|---|---|
| hydraulic | adj. | 水力的，水压的 |
| hydroelectric | adj. | 水电的，水力电气的 |
| hydroenergy | n. | 水能 |
| hydropower | n. | 水电，水力发出的电力 |
| hydromechanics | n. | 流体力学 |
| hydrovalve | n. | 水龙头，水阀，液压开关 |

## 2. thermo- 表示：热，热电

| | | |
|---|---|---|
| thermoammeter | n. | （测量微电流用的）热电偶安培计 |
| thermoanalysis | n. | 热（学）分析 |
| thermobattery | n. | 温差电池（组），热电池（组） |
| thermocell | n. | 温差电偶 |
| thermochemical | adj. | 热化学的 |
| thermocontact | n. | 热接触 |
| thermocurrent | n. | 热电流 |
| thermocutout | n. | 热保险装置[断流器] |
| thermoelectric | adj. | 热电的 |
| thermoelement | n. | 热电偶，热电元件 |
| thermostat | n. | 温度调节装置 |

# Unit 1

# Review

## I. Summary of Glossaries: Power Generation

1. 发电厂
   - power plant　　　　　　　　　　发电厂
   - power station　　　　　　　　　发电厂，电站
   - power house　　　　　　　　　　发电站，动力车间
2. 水电站
   - hydropower station/ plant　　　水力发电厂
   - hydroelectric power station　　水力发电站
   - water power station　　　　　　水电厂
3. 热电厂，火电厂
   - thermal plant　　　　　　　　　热电厂
   - heat-engine plant　　　　　　　热电厂
   - fossil-fuel plant　　　　　　　化石燃料电厂
   - coal-fired plant　　　　　　　　燃煤电厂
   - steam-turbine plant　　　　　　蒸汽轮机电厂
   - gas-turbine plant　　　　　　　燃气轮机电厂
4. **fossil fuel**　　　　　　　　　　**化石燃料**
   - coal　　　　　　　　　　　　　　煤，煤炭
   - oil, petroleum　　　　　　　　　油，石油
   - gas, natural gas　　　　　　　　气，天然气
5. **nuclear reaction**　　　　　　　**核反应**
   - nuclear station　　　　　　　　核电站
   - nuclear reaction　　　　　　　　核反应
   - nuclear reactor　　　　　　　　核反应堆
   - nuclear fission　　　　　　　　核裂变
   - nuclear fusion　　　　　　　　　核聚变

## II. Abbreviations (Abbr.)

1. DC　　direct current　　　　　直流（电）
2. AC　　alternating current　　交流（电）

# Special Topic (1)　　Basic Electric Quantities

There are many electric quantities used for describing the state or nature of a power system. Some can be measured directly by instruments, and others can be calculated due to the measured data.

The common electric quantities often used and their basic units are shown in Table 1.

Table 1　Basic electric quantities and their unit

| | Quantity | | | Unit | |
| --- | --- | --- | --- | --- | --- |
| Symbol | name | meaning | Symbol | name | meaning |
| $V$ | Voltage | 电压 | V | Volt | 伏，伏特 |
| $I$ | Current | 电流 | A | Ampere | 安，安培 |
| $f$ | Frequency | 频率 | Hz | Hertz | 赫，赫兹 |
| $T$ | Period | 周期 | s | second | 秒 |
| $P$ | Active power | 有功功率 | W | Watt | 瓦，瓦特 |
| $Q$ | Reactive power | 无功功率 | Var | Var | 乏，乏尔 |
| $S$ | Apparent power | 视在功率 | VA | VA | 伏安 |

# Reading Material —— The One-Line Diagram

Since a balanced three-phase system is always solved as a single-phase circuit composed of one of the three lines and a neutral return, it is seldom necessary to show more than one phase and the neutral return when drawing a diagram of the circuit. Often the diagram is simplified further by omitting the complete circuit through the neutral and by indicating the component parts by standard symbols rather than by their equivalent circuits.* Circuit parameters are not shown, and the transmission line is represented by a single line between its two ends. Such a simplified diagram of an electric system is called a one-line diagram. It indicates by a single line and standard symbols the transmission lines and associated apparatus of an electric system.

The purpose of the one-line diagram is to supply in concise form significant information about the system.* The importance of different features of a system varies with the problem under consideration, and the amount of information included on the diagram depends on the purpose for which the diagram is intended.* For instance, the location of circuit breakers and relays is unimportant in making a load study. Breakers and relays are not shown if the primary function of the diagram is to provide information for such a study. On the other hand, the determination of the stability of a system under transient conditions resulting from a fault depends on the speed with which relays and circuit breakers operate to isolate the faulted part of the system.* Therefore, information about the circuit breakers may be of extreme importance. Sometimes one-line diagrams include information about the current and potential transformers which connect the relays to the system or which are installed for metering. The information found on a one-line diagram must be expected to vary according to the problem at hand and according to the practice of the particular company preparing the diagram.*

The American National Standards Institute (ANSI) and the Institute of Electrical and Electronics Engineers (IEEE) have published a set of standard symbols for electrical diagrams.

## New Words and Expressions

| | | | |
|---|---|---|---|
| 1. | one-line diagram | n. | 单线电路图，单线图 |
| 2. | balanced | adj. | 平衡的 |
| 3. | neutral return | | 中性线回路 |
| 4. | omit | vt. | 省略，疏忽，遗漏 |
| 5. | component part | | （组）成（部）分 |
| 6. | parameter | n. | 参数，参量 |
| 7. | transmission line | | 传输线，输电线 |
| 8. | associated apparatus | | 相关设备 |
| 9. | concise | adj. | 简明的，简练的 |
| 10. | intend for | v. | 打算供…使用 |
| 11. | circuit breaker | | 断路开关，断路器 |
| 12. | relay | n. | 继电器 |
| 13. | primary function | | 主要功能；基函数，原函数 |
| 14. | stability | n. | 稳定性 |
| 15. | transient | n. | 瞬时现象，过渡过程 |
| | | adj. | 短暂的，瞬时的 |
| 16. | fault | n. | 故障 |
| 17. | potential transformer | | 电压互感器 <PT> |
| 18. | meter (metering) | vt. | 测量（法），计[配]量，测定 |
| 19. | at hand | adv. | 在手边，在附近，即将到来 |

## Abbreviations

| | | |
|---|---|---|
| 1. | ANSI | the American National Standards Institute 美国国家标准学会 |
| 2. | IEEE | the Institute of Electrical and Electronics Engineers 电力与电子工程师协会 |

## Notes

**1. Often the diagram is simplified further by omitting the complete circuit through the neutral and by indicating the component parts by standard symbols rather than by their equivalent circuits.**

译文：单线图往往被进一步简化，简化的途径包括：忽略经过中线的完整电路，用标准符号表示电路组成部分而不是其等效电路。

说明：状语 often 放在句首是为了强调这种情况很常见，按照中文表达习惯，这种表示频度的状语一般应放在主语后面。句子的主干很短，而由 by 引导的两个状语短语则较长，若直译则显得很拗口。遇到这种情况，可以进行一定的变化，如上述译文所示。为便于理解，句子可以拆解为：

The diagram is often simplified further by some means; these means include: (1) omitting the complete circuit through the neutral, and (2) indicating the component parts by standard symbols rather than by their equivalent circuits.

**2. The purpose of the one-line diagram is to supply in concise form significant information about the system.**

译文：单线图的用途是以简单明了的形式提供关于电力系统的重要信息。

说明：这是一个单句，结构不是很复杂。需要注意的是，状语短语 in concise form 放在了宾语短语 significant information 的前面，容易造成歧义或者混淆。为便于理解，句子可以调整为：

The purpose of the one-line diagram is to supply significant information about the system in concise form.

**3. The importance of different features of a system varies with the problem under consideration, and the amount of information included on the diagram depends on the purpose for which the diagram is intended.**

译文：系统不同特征的重要性随着所考虑的问题不同而变化，单线图所包含的信息量取决于它用于什么目的。

说明：从句中出现了一个拆开的词组 intend for，意思为"打算供……使用"。定语从句 for which the diagram is intended 用于修饰 the purpose。为了便于理解，这个句子可以拆解为：

The importance of different features of a system varies with the problem under consideration, and the amount of information included on the diagram depends on the purpose; the purpose is which the diagram is intended for.

**4. On the other hand, the determination of the stability of a system under transient conditions resulting from a fault depends on the speed with which relays and circuit breakers operate to isolate the faulted part of the system.**

译文：另一方面，在由故障导致的暂态条件下，系统稳定性的确定取决于继电器和断路器将故障部分从系统中隔离的速度。

说明：On the other hand 意思是"另一方面"。with which 引导的从句作为 the speed 的定语。为了便于理解，这个句子可以拆解为：

On the other hand, the determination of the stability of a system under transient conditions resulting from a fault depends on the speed; relays and circuit breakers operate to isolate the faulted part of the system with the speed.

**5. The information found on a one-line diagram must be expected to vary according to the problem at hand and according to the practice of the particular company preparing the diagram.**

译文：一定期望基于单线图（得到的）信息随着手边的问题和准备单线图的特定公司

的实际情况而变化。

说明：动词 vary 后面有两个并列的由 according to 引导的状语短语。分词短语 preparing the diagram 是修饰 the particular company 的定语。为了便于理解，这个句子可以拆解为：

The information found on a one-line diagram must be expected to vary according to some factors; these factors are the problem at hand and the practice of the particular company; the company is who prepares the diagram.

# Knowledge & Skills (1)　Parenthesis

**1. 插入语的概念**

A qualifying or amplifying word, phrase, or sentence inserted within written matter in such a way as to be independent of the surrounding grammatical structure.

插入语：在书写中插入的与上下文的语法结构没有关系的限定或补充的词、短语或句子。

**2. 插入语的特点**

- 表示时间、条件、场合等的状语短语，用来明确或强调所述情况成立的前提条件。
- 常见的有：
  - as shown in Figure.x (or Table)　如图（表）x 所示
  - for instance　例如
  - under ... condition　在……条件下
  - in recent years　近年来
- 在翻译时，应按汉语习惯，一般应提前到句首。

**3. 举例**

例 1：The industry, by some standards, is the largest in the world.

按照某种标准，该行业是世界上最大的。

说明：by some standards 是插入语，意思是"按照某种标准，依据特定的标准"。在翻译时其对应内容应提到句首。

例 2：Even though copper is, under present market conditions, more expensive than aluminum, the savings in the cost of insulation and the increased flexibility of smaller diameters make copper preferable to aluminum in the design of most high-voltage power cables.

尽管在目前的市场行情中，铜的价格比铝贵，但节省下来的用于绝缘的费用，以及直径小带来的可挠曲性提高，使得在大多数高压电力电缆的设计中铜比铝更优越。

说明：in local regions 是插入语，意思是"在局部区域"。在翻译时其对应内容应提到句首。

例 3：Such changes may, in local regions, result in the insulating material becoming highly conductive.

在局部区域，这种变化会导致绝缘材料变得高度导电。

说明：in local regions 是插入语，意思是"在局部区域"。在翻译时其对应内容应提到句首。

例4：The power system, as shown in Figure 1, consists of three main parts.

如图1所示，电力系统由三部分组成。

说明：as shown in Figure 1 是插入语，意思是"如图1所示"。在翻译时其对应内容应提到句首。

例5：FACTS, in recent years, has more and more applications.

近年来，FACTS 有越来越多的应用。

说明：in recent years 是插入语，意思是"近年来"。在翻译时其对应内容应提到句首。

# Chapter 4
# Transmission System and Transmission Lines

## Part 1  General Introduction of the Transmission System

The transmission system interconnects all major generating stations and main load centers in the system. It forms the backbone of the integrated power system and operates at the highest voltage levels (typically, 230kV and above). The generator voltages are usually in the range of 11 to 35kV. These are stepped up to the transmission voltage level, and power is transmitted to transmission substations where the voltages are stepped down to the sub-transmission level (typically, 69 to 138kV). The generation and transmission subsystems are often referred to as the *bulk power system.*

The sub-transmission system transmits power in smaller quantities from the transmission substations to the distribution substations. Large industrial customers are commonly supplied directly from the sub-transmission system. In some systems, there is no clear demarcation between sub-transmission and transmission circuits. As the system expands and higher voltage levels become necessary for transmission, the older transmission lines are often relegated to sub-transmission function.*

Electric power is transferred from generating stations to consumers through overhead lines and cables. Although these conductors appear very ordinary, they possess important electrical properties that greatly affect the transmission of electrical energy.

## Part 2  Considerations for Transmission Lines and Cables

Overhead Lines are used for long distances in open country and rural areas, whereas cables are used for underground transmission in urban areas and for underwater crossings. For the same rating, cables are 10 ~ 15 times more expensive than overhead lines and they are therefore only used in special situations where overhead lines cannot be used.

The advantage of a higher level of transmission-line voltage is apparent when consideration is given to the transmission capability in megavoltamperes (MVA) of a line. Voltage regulation, thermal limit, and system stability are the factors that determine the power transmission capability of power lines. Roughly, the capability of lines of the same length varies <u>at a rate</u> somewhat greater than the square of the voltage. No definite capability can be specified for a line of a given voltage, however, because capability is dependent on the thermal limits of the conductor, allowable voltage drop, reliability, and requirements for maintaining synchronism between the machines of the system, which is known as stability.* Most these factors are dependent on line length. <u>The greater</u> the required capacity or line length, <u>the higher</u> the optimum voltage class.

The heat produced by current flow in transmission lines has two undesirable effects: ① Annealing and gradual loss of mechanical strength of the aluminum conductor caused by continued <u>exposure to</u> temperature extremes; ② Increased sag and decreased clearance to the ground due to conductor expansion at higher temperatures. The second of the above two effects is generally the limiting factor in setting the maximum permissible operating temperature. At this limit, the resulting line sag approaches the statutory minimum ground clearance. The maximum allowable conductor temperatures based on annealing considerations are 127℃ for conductors with high aluminum content and 150℃ for other conductors. The allowable maximum current (i.e., the ampacity) depends on the ambient temperature and the wind velocity. Lines have summer and winter emergency ratings respectively. These are design values based on extreme values of ambient temperature, wind velocity, and solar radiation.

Other considerations for an overhead line include: the number, size, and spacing of conductors per-phase bundle; the phase-to-phase spacing; the number, location, and conductor type for overhead neutrals concerning lightning shielding; the level of insulation; etc.

The design of a power line depends upon the following criteria:

1) The amount of active power it has to transmit.
2) The distance over which the power must be carried.
3) The cost of the power line.
4) Esthetic considerations, urban congestion, ease of installation, and expected load growth.

## Part 3  Circuit Parameters of Transmission Lines and Cables

An electric transmission line has four parameters that affect its ability to fulfill its function as part of a power system: resistance, inductance, capacitance, and conductance.

Conductance between conductors or between conductors and the ground accounts for the leakage current at the insulators of overhead lines and through the insulation of cables. Since leakage at insulators of overhead lines is negligible, the conductance between conductors of an overhead line is assumed to be zero.

Capacitance exists between the conductors and is the charge on the conductors per unit of potential difference between them.*

# Chapter 4  Transmission System and Transmission Lines

The resistance and inductance uniformly distributed along the line form the series impedance. The conductance and capacitance existing between conductors of a single-phase line or from a conductor to neutral of a three-phase line form the shunt admittance. Although the resistance, inductance, and capacitance are distributed, the equivalent circuit of a line is made up of lumped parameters.

Uniform distribution of current throughout the cross-section of a conductor exists only for direct current. As the frequency of alternating current increases the non-uniformity of distribution becomes more pronounced. An increase in frequency causes non-uniform current density. This phenomenon is called the *skin effect*. In a circular conductor, the current density *usually* increases from the interior toward the surface. For conductors of sufficiently large radius, however, a current density oscillation with respect to radial distance from the center may result.

When the conductors of a three-phase line are not spaced equilaterally, the problem of finding the inductance becomes more difficult. Then the flux linkages and inductance of each phase are not the same. A different inductance in each phase results in an unbalanced circuit. Balance of the three phases can be restored by exchanging the positions of the conductors at regular intervals along the line so that each conductor occupies the original position of every other conductor over an equal distance, as shown in Figure 4.1. *Transposition results in each conductor having the same average inductance over the whole cycle.

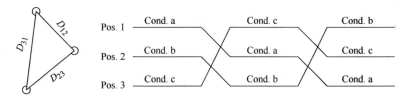

**Figure 4.1**  Transposition Cycle

## Part 4  Types of Transmission Conductors

In the early days of the transmission of electric power, conductors were usually copper, but aluminum conductors have completely replaced copper because of the much lower cost and lighter weight of an aluminum conductor compared with a copper conductor of the same resistance.* The fact that an aluminum conductor has a larger diameter than a copper conductor of the same resistance is also an advantage. With a larger diameter, the lines of electric flux originating on the conductor will be farther apart at the conductor surface for the same voltage. This means a lower voltage gradient at the conductor surface and less tendency to ionize the air around the conductor. Ionization produces the undesirable effect called *corona*.

Symbols identifying different types of aluminum conductors are as follows:

| | |
|---|---|
| AAC | all-aluminum conductors |
| AAAC | all-aluminum-alloy conductors |
| ACSR | aluminum conductor, steel-reinforced |
| ACAR | aluminum conductor, alloy-reinforced |

Aluminum-alloy conductors have higher tensile strength than the ordinary electrical-conductor grade of aluminum. ACSR consists of a central core of steel strands surrounded by layers of aluminum strands. ACAR has a central core of higher-strength aluminum surrounded by layers of electrical-conductor-grade aluminum.

Alternate layers of wire of a stranded conductor are spiraled in opposite directions to prevent unwinding and make the outer radius of one layer coincide with the inner radius of the next. The number of strands depends on the number of layers and on whether all the strands are the same diameter. The total number of strands in concentrically stranded cables, where the total annular space is filled with strands of uniform diameter, is 7, 19, 37, 61, 91, or more.*

Cables for underground transmission are usually made with stranded copper conductors rather than aluminum. The conductors are insulated with oil-impregnated paper. Up to a voltage of 46kV, the cables are of the solid type which means that the only insulating oil in the cable is that which is impregnated during manufacture.*

## Part 5   Power Transmission Towers

Pole and Tower is a pole-type or tower-type structure that supports overhead transmission line conductors and overhead ground wires and keeps a certain distance between them and the earth.

The classification of the pole tower:

### 1. According to the classification of structural materials

- Wood structure
- Steel structure
- Aluminum alloy structure
- Reinforced concrete structure

### 2. According to the classification of structure form

- Self-supporting support: Without cable, the tower itself has stability.
- Guyed tower: Overhead transmission lines to support the support structure of conductors and lightning conductors.

### 3. According to the classification of the use function

- Bearing tower: It is set up in sections, the tension insulator string of the conductor is anchored on the tower to bear the hanging tension of wires and ground wires on both sides

# Chapter 4　Transmission System and Transmission Lines

and the unbalanced tension during the accident. Bearing towers can also <u>be divided into</u> strain towers, angle towers, and terminal towers.

- Tangent tower: It is generally used to bear the gravity of the wire.
- Transposition tower: Achieve wire transposition.
- Large crossing tower: Refers to the large span high tower that spans a navigable river.

**4. According to the classification of the same tower erected transmission line circuit number**

- Single-loop tower
- Dual-loop tower
- Multi-loop tower

## NEW WORDS AND EXPRESSIONS

| | | | |
|---|---|---|---|
| 1. | backbone | *n.* | 中枢 |
| 2. | integrated | *adj.* | 综合的，完整的 |
| 3. | voltage level | | 电压等级 |
| 4. | step up | | 提升，提高，升压 |
| 5. | step down | | 减低，降压 |
| 6. | *bulk* | *n.* | 主要部分，大量 |
| | | *adj.* | 大批的，大量的 |
| 7. | demarcation | *n.* | 划分 |
| 8. | overhead line | | 架空线 |
| 9. | cable | *n.* | 电缆 |
| 10. | megavoltampere | *n.* | 兆伏安 <MVA> |
| 11. | thermal limit | | 发热限制，发热极限 |
| 12. | voltage drop | | 电压降落 |
| 13. | synchronism | *n.* | 同步性，同时性 |
| 14. | optimum | *a.n.* | 最适宜（的） |
| 15. | voltage class | | 电压等级 |
| 16. | annealing | *v.* | 退火 |
| 17. | mechanical strength | | 机械强度 |
| 18. | sag | *n.* | 下陷；暂降 |
| 19. | clearance | *n.* | 间隔，距离 |
| 20. | statutory minimum | | 法定最小限度 |
| 21. | content | *n.* | 比例 |
| 22. | ampacity | *n.* | 安培容量，载流容量 |
| 23. | ambient | *adj.* | 周围的 |
| 24. | solar radiation | | 太阳辐射 |
| 25. | bundle | *n.* | 捆，束，包 |
| 26. | neutral | *n.* | 中性点 |
| 27. | lightning shielding | | 雷电防护 |
| 28. | resistance | *n.* | 电阻 |

| | | | |
|---|---|---|---|
| 29. | inductance | n. | 电感 |
| 30. | capacitance | n. | 电容 |
| 31. | conductance | n. | 电导 |
| 32. | leakage | n. | 漏，泄漏 |
| 33. | insulator | n. | 绝缘体，绝热器 |
| 34. | impedance | n. | 阻抗，全电阻 |
| 35. | shunt | n. | 旁路，并联 |
| 36. | admittance | n. | 导纳 |
| 37. | equivalent circuit | | 等效电路 |
| 38. | lumped parameter | | 集中参数，集总参数 |
| 39. | uniform distribution | | 均匀分布 |
| 40. | non-uniformity | n. | 不均匀 |
| 41. | pronounced | adj. | 显著的，断然的，明确的 |
| 42. | skin effect | | 集肤[趋肤]效应 |
| 43. | oscillation | n. | 振荡，上下波动 |
| 44. | equilaterally | adv | 等边地 |
| 45. | flux linkage | | 磁通匝连数，磁链 |
| 46. | copper | n. | 铜 |
| 47. | electric flux | | 电通，电通量 |
| 48. | gradient | n. | 梯度，坡度 |
| 49. | tendency | n. | 趋向，倾向 |
| 50. | ionize | v. | 电离，使离子化 |
| 51. | corona | n. | 电晕 |
| 52. | alloy | n. | 合金 |
| 53. | tensile | adj. | 张力的，拉力的 |
| 54. | strand | n. | 绳，线；绳、线之一股 |
| 55. | alternate layers | | 轮换层 |
| 56. | spiral | v. | 螺旋形盘绕 |
| 57. | unwind | vt. | 解开，松开 |
| 58. | concentrically | adj. | 集中的，共中心的，同轴的 |
| 59. | annular space | | 环形空隙[空间] |
| 60. | impregnate | vt. | 注入，使充满 |
| 61. | aluminum | n. | 铝 |
| 62. | self-supporting tower | | 自立塔 |
| 63. | guyed tower | | 拉线塔 |
| 64. | bearing tower | | 承力塔 |
| 65. | insulator | n. | 绝缘体，绝缘子 |
| 66. | anchor | v. | 抛锚，使固定 |
| 67. | strain tower | | 耐张塔 |
| 68. | angle tower | | 转角塔 |
| 69. | terminal tower | | 终端塔 |
| 70. | gravity | n. | 重力 |
| 71. | tangent tower | | 直线塔 |
| 72. | transposition tower | | 换位塔 |

# Chapter 4　Transmission System and Transmission Lines

## PHRASES

**1. refer to ... as:** 称（某人，某事物）…为

Example: The generation and transmission subsystems are often referred to as the bulk power system.

发电和输电子系统常常被称为大系统。

**2. relegate to:** 移交给，推给

Example: As the system expands and higher voltage levels become necessary for transmission, the older transmission lines are often relegated to sub-transmission function.

随着系统的扩大和更高的电压等级为输电所必需，旧的输电线路往往被移交给中压输电功能。

**3. n times more (expensive) than b:** 比 b 多（贵）n 倍

Example: For the same rating, cables are 10～15 times more expensive than over-head lines and they are therefore only used in special situations where overhead lines cannot be used.

在相同的额定情况下，电缆比架空线贵 10～15 倍，它们只在不能用架空线的场合使用。

**4. at a rate:** 以……速度

Example: Roughly, the capability of lines of the same length varies at a rate somewhat greater than the square of the voltage.

粗略来说，同样长度的线路的容量变化趋势快于电压二次平方。

**5. the greater..., the higher...:** 越……越……

Example: The greater the required capacity or line length, the higher the optimum voltage class.

需要的容量越大，线路越长，则最佳的电压等级越高。

**6. expose to:** 使易受，使受

Example: Annealing and gradual loss of mechanical strength of the aluminum conductor caused by continued exposure to temperature extremes.

持续经受高温极端条件所引起的退火和机械强度的逐渐丧失。

**7. with respect to:** 由于，与……有关

Example: For conductors of sufficiently large radius, however, a current density oscillation with respect to radial distance from the center may result.

然而，对半径足够大的导线而言，可能导致与距导线中心的径向距离有关的电流密度振荡。

**8. coincide with:** 与……一致

Example: Alternate layers of wire of a stranded conductor are spiraled in opposite directions to prevent unwinding and make the outer radius of one layer <u>coincide with</u> the inner radius of the next.

绞线的导线轮换层从相反的方向螺旋缠绕以防松开，并且使某一层的外边界与邻近层的内边界一致。

**9. be divided into: 被分成**

Example: Bearing tower can also <u>be divided into</u> strain tower, angle tower and terminal tower.

承力塔可以被分成耐张塔、转角塔和终端塔。

## COMPLICATED SENTENCES

**1. As the system expands and higher voltage levels become necessary for transmission, the older transmission lines are often <u>relegated to</u> sub-transmission function.**

译文：随着系统的扩大和更高的电压等级为输电所必需，旧的输电线路往往被移交给中压输电功能。

说明：higher voltage levels become necessary for transmission 与 system expands 并列，在翻译时要尽量避免两部分的字数差别太大。be relegated to 表示"移交给……，推给……"。

**2. No definite capability can be specified for a line of a given voltage, however, because capability <u>is dependent on</u> the thermal limits of the conductor, allowable voltage drop, reliability, and requirements for maintaining synchronism between the machines of the system, which <u>is known as</u> stability.**

译文：然而，无法为给定电压等级的线路指定确切的容量，这是因为容量取决于导线的发热极限、允许的电压降落、可靠性和维持系统中的发电机同步的要求（这被认为是稳定性）。

说明：however 是承接句子前面的其他内容而言的，在翻译时应该提到句首。be dependent on 相当于 depend on，表示"依赖于，取决于"。be known as 意思是"被认为是……"。句中 be dependent on 后面带了多个宾语，其中最后一个宾语还有自己的定语从句，为了不影响这些宾语的并列陈述，宾语的定语从句可以以括号内容的形式出现。Which 引导的定语从句修饰的主体是 synchronism between the machines of the system。为了便于理解，这个句子可以拆解为：

However, No definite capability can be specified for a line of a given voltage, because capability is dependent on the thermal limits of the conductor, allowable voltage drop, reliability, and requirements for maintaining synchronism between the machines of the system; synchronism between the machines is known as stability.

**3. Capacitance exists between the conductors and is the charge on the conductors per unit of potential difference between them.**

译文：电容存在于导体之间，等于导体上的电荷（电量）与导体间的电势差之比。

# Chapter 4 Transmission System and Transmission Lines

或：电容存在于导体之间，等于导体之间单位电压对应的电量。

说明：这个句子结构并不复杂，翻译的难点在于 per unit of 如何翻译。A per unit of B 表示"单位 B 上的 A"或者"A/B"。

**4. Balance of the three phases can be restored by exchanging the positions of the conductors at regular intervals along the line so that each conductor occupies the original position of every other conductor over an equal distance.**

译文：沿线路方向每隔一定间隔就交换导线位置，使每根导线都能占据具有相同输电距离的其他各导线的初始位置，以重建三相平衡。

说明：句子的主语和谓语主干很短，但带有一个很长的状语，即从 by 开始到句子结尾的全部内容。这个状语短语的结构比较复杂。由 so that 引导的从句，说明 exchanging the positions 的目的，over an equal distance 是 every other conductor 的定语。为了便于理解，这个句子可以拆解为：

Balance of the three phases can be restored by exchanging the positions of the conductors at regular intervals along the line; exchanging the positions of the conductors so that each conductor occupies the original position of every other conductor; the conductor and every other conductor are over an equal distance.

**5. Aluminum conductors have completely replaced copper because of the much lower cost and lighter weight of an aluminum conductor compared with a copper conductor of the same resistance.**

译文：与具有相同电阻的铜导线相比，铝的成本更低，重量更轻，铝导线已经完全取代了铜导线。

说明：because of 后面接带定语的名词短语时，该名词短语往往不宜直接翻译，很多时候可以处理为主谓结构的短语。例如，the much lower cost and lighter weight of an aluminum conductor 可以翻译为：铝的成本更低，重量更轻。

**6. The total number of strands in concentrically stranded cables, where the total annular space is filled with strands of uniform diameter, is 7, 19, 37, 61, 91, or more.**

译文：同心绞合电缆内的所有的环形空间都被直径相同的绞线填满，绞线的总股数为 7、19、37、61、91 或者更多。

说明：这是 where 引导的定语从句，修饰的主体是 stranded cables，该从句插入到了句子中间，并以逗号前后分隔，容易干扰视线，翻译时要弄清句子结构。另外，句子的谓语是从 is 开始的，比较短，而主语带了一个很长的修饰语，而且该修饰语自己还嵌套有定语从句。为了避免"头重脚轻"，翻译时可以对句子的顺序和结构作适当调整。为了便于理解，这个句子可以拆解为：

The total number of strands in concentrically stranded cables is 7, 19, 37, 61, 91, or more; in the concentrically stranded cables, the total annular space is filled with strands of uniform diameter.

**7. Up to a voltage of 46kV the cables are of the solid type which means that the only insulating oil in the cable is that which is impregnated during manufacture.**

译文：到 46kV 电压，电缆为固态类型，即电缆中仅有的绝缘油是在制造时注入的。

说明：句子中第一个 which 引导的定语从句修饰的主体是 solid type，而该定语从句内部还嵌套了一个 which 引导的定语从句，修饰的主体是 that。其实这个句子可以简化为：

Up to a voltage of 46kV the cables are of the solid type; the solid type means that the only insulating oil in the cable is impregnated during manufacture.

## ABBREVIATIONS (ABBR.)

| | | | |
|---|---|---|---|
| 1. | MVA | megavoltamperes | 兆伏安 |
| 2. | kV | kilovoltage | 千伏电压 |
| 3. | AAC | all-aluminum conductor | 全铝导体 |
| 4. | AAAC | all-aluminum-Alloy Conductor | 全铝合金导体 |
| 5. | ACSR | aluminum conductor, steel-reinforced | 铝导体，钢加固 |
| 6. | ACAR | aluminum conductor, alloy-reinforced | 铝导体，合金加固 |

## SUMMARY OF GLOSSARY

1. 电压等级变化
   - step up　　　　　　　　　　　　　　升压
   - step down　　　　　　　　　　　　　降压
2. 输电系统
   - transmission subsystem　　　　　　　输电子系统
   - sub-transmission system　　　　　　 中（等电）压输电系统
3. 变电站
   - transmission substation　　　　　　　输电变电站
   - distribution substation　　　　　　　 配电变电站
4. 输电线
   - transmission line　　　　　　　　　　输电线，传输线
   - overhead line　　　　　　　　　　　 架空线
   - power line　　　　　　　　　　　　　电力线
   - cable　　　　　　　　　　　　　　　 电缆
5. 线路参数 circuit parameters
   - resistance　　　　　　　　　　　　　电阻
   - inductance　　　　　　　　　　　　　电感
   - capacitance　　　　　　　　　　　　 电容
   - conductance　　　　　　　　　　　　电导
   - impedance　　　　　　　　　　　　　阻抗
   - admittance　　　　　　　　　　　　　导纳

6. 电路元件的基本连接方式
series 串联
shunt 并联，旁路

# EXERCISES

**1. Translate the following words or expressions into Chinese.**

(1) voltage level  (2) subsystem  (3) thermal limit
(4) overhead line  (5) clearance  (6) megavoltampere
(7) shunt admittance  (8) skin effect  (9) mechanical strength
(10) unwind  (11) equivalent circuit  (12) angle tower

**2. Translate the following words or expressions into English.**

(1) 中性点  (2) 电缆  (3) 磁链
(4) 绝缘  (5) 同步性  (6) 串联
(7) 横截面  (8) 电晕  (9) 重力

**3. Fill in the blanks with proper words or expressions.**

(1) Electric power is transferred from generating stations to consumers through _____ and _____.

(2) Overhead Lines are used for _____ distances in open country and _____ areas, whereas cables are used for _____ transmission in _____ areas and for _____ crossings.

(3) An electric transmission line has four parameters that affect its ability to fulfill its function as part of a power system: _____, _____, _____, and _____, which may form series _____ and shunt _____.

# Word-Building (4)  后缀 -or, er; -ance  机，器，装置，元件

**1. -er, or [名词后缀]，表示：…机，…器，…装置**

| | | | | |
|---|---|---|---|---|
| generate | vt. | generator | n. | 发电机 |
| compress | vt. | compressor | n. | 压缩机 |
| motion | v. | motor | n. | 发动机，电动机 |
| regulate | vt. | regulator | n. | 调整器，调节器 |
| compensate | v. | compensator | n. | 补偿器 |
| control | n. | controller | n. | 控制器 |
| excite | v. | exciter | n. | 励磁器 |
| condition | vt. | conditioner | n. | 调节装置 |

| | | | | | |
|---|---|---|---|---|---|
| reclose | vt. | recloser | n. | 自动重合闸装置 | |
| rotate | v. | rotor | n. | 转子 | |
| static | adj. | stator | n. | 定子 | |

## 2. -er,or [名词后缀]，表示：…元件，…器

| | | | | |
|---|---|---|---|---|
| conduct | vt. | conductor | n. | （电）导体 |
| capacity | n. | capacitor | n. | 电容器 |
| react | vi. | reactor | n. | 电抗器 |
| resist | vt. | resistor | n. | 电阻器，电阻元件 |
| induct | v. | inductor | n. | 电感元件 |

## 3. -ance/-ence [名词词尾]，表示某种物理量，或相关系数

| | | |
|---|---|---|
| resistance | n. | 电阻，阻抗 |
| conductance | n. | 电导 |
| capacitance | n. | 容量，电容 |
| inductance | n. | 电感；感应；感应系数 |
| reactance | n. | 电抗 |
| impedance | n. | 阻抗，全电阻 |
| susceptance | n. | 电纳 |
| admittance | n. | 导纳 |

# Chapter 5
# Power Transformers

## Part 1  Power Transformers Used in Power System

The transformer is probably one of the most useful electrical devices ever invented. An electric transformer is a device that changes AC electric energy at one voltage level into AC electric energy at another voltage level through the action of a magnetic field.* It can raise or lower the voltage or current in an AC circuit, it can isolate circuits from each other, and it can increase or decrease the apparent value of a capacitor, an inductor, or a resistor. Furthermore, the transformer enables us to transmit electrical energy over great distances and to distribute it safely in factories and homes.

From the viewpoints of efficiency and power-transfer capability, the transmission voltages have to be high, but it is not practically feasible to generate and consume power at these voltages. Transformers enable the utilization of different voltage levels across the system. Generator voltage, usually in the range of 13.8kV to 24kV, is stepped up to transmission level in the range of 115kV to 765kV by a *unit transformer* which is connected to the output of the generator. The first step-down of voltage from transmission levels to distribution levels is at the bulk-power substation, where the reduction is to a range of 34.5 to 138kV, depending, of course, upon transmission-line voltage.* The transformers used here are called *substation transformers*. Some industrial customers may be supplied at these voltage levels. The next step-down in voltage is at the distribution system using a *distribution transformer* for most customers to use. There are some *interconnecting transformers* between different power systems that are interconnected. All these transformers are essentially the same devices used in different situations.

In modern electric power systems, the transmitted power undergoes four to five voltage transformations between the generators and the ultimate consumers. Consequently, the total MVA rating of all the transformers in a power system is about five times the total MVA ratings of all the generators.

## Part 2  The Structure of a Transformer

A transformer is a device that consists of two or more coils placed so that they are linked by the same flux. In a power transformer the coils are placed on an iron core which is made of ferromagnetic material in order to confine the flux so that almost all of the flux linking any one coil links all the others.* Several coils may be connected in series or parallel to form one winding, the coils of which may be stacked on the core alternately with those of the other winding or windings. The number of turns in a winding may be several hundred up to several thousand.

The winding across which impedance or other load may be connected is called the *secondary* winding or output winding and any circuit elements connected to this winding are said to be on the secondary side of the transformer.* Similarly the winding which is toward the source of energy is called the *primary* winding or input winding on the primary side. In the power system, energy often will flow in either direction through a transformer and the designation of primary and secondary loses its meaning. If there is a third winding on the transformer, it is called the *tertiary* winding.

An autotransformer differs from the ordinary transformer in that the windings of the autotransformer are electrically connected as well as being coupled by a mutual flux, as will be described later.

Three identical single-phase transformers may be connected so that the three windings of one voltage rating are $\Delta$-connected and the three windings of the other voltage rating are Y-connected to form a three-phase transformer. Such a transformer is said to be connected Y-$\Delta$ or $\Delta$-Y. The other possible connections are Y-Y and $\Delta$-$\Delta$. If the three single-phase transformers each have three windings (primary, secondary, and tertiary), two sets might be connected in Y and one in $\Delta$, or two could be $\Delta$-connected with one Y-connected. Rather than use three identical single-phase transformers, a more usual unit is a three-phase transformer where all three phases are on one iron structure. The theory is the same for a three-phase transformer as for a three-phase bank of single-phase transformers.

## Part 3　The Ideal Transformer

We shall begin our analysis by assuming that the flux varies sinusoidally in the core and that the transformer is *ideal*, which means that the permeability $\mu$ of the core is infinite and the resistance of the windings is zero.* With infinite permeability of the core all of the flux is confined to the core and therefore links all of the turns of both windings. The voltage $e$ induced in each winding by the changing flux is also the terminal voltage $v$ of the winding since the winding resistance is zero.

Voltage $e_1$ and $e_2$ induced by the changing flux are in phase when they are defined in certain directions. Then by Faraday's law

$$v_1 = e_1 = N_1 \frac{d\phi}{dt} \quad v_2 = e_2 = N_2 \frac{d\phi}{dt} \tag{5-1}$$

where $\Phi$ is the instantaneous value of the flux and $N_1$ and $N_2$ are the number of turns on windings 1 and 2. Since we have assumed a sinusoidal variation of the flux we can convert to phasor form after dividing the above equations to yield

# Chapter 5 Power Transformers

$$\frac{\dot{V}_1}{\dot{V}_2} = \frac{\dot{E}_1}{\dot{E}_2} = \frac{N_1}{N_2} \tag{5-2}$$

Usually, we do not know the direction in which the coils of a transformer are wound. One device to provide winding information is to place a dot at the end of each winding such that all dotted ends of windings are positive at the same time; that is, voltage drops from dotted to unmarked terminals of all windings are in phase.* So

$$\frac{\dot{I}_1}{\dot{I}_2} = \frac{N_1}{N_2} \tag{5-3}$$

Note then that $I_1$ and $I_2$ are in phase if we choose the current to be positive when entering the dotted terminal of one winding and leaving the dotted terminal of the other.

If an impedance $Z_2$ is connected across winding 2

$$Z_2 = \frac{\dot{V}_2}{\dot{I}_2} \tag{5-4}$$

but upon substituting for $V_2$ and $I_2$ values determined from Eqs. (5-3) and (5-4), the impedance, as measured across the primary winding, is

$$Z_2' = \frac{\dot{V}_1}{\dot{I}_1} = \left(\frac{N_1}{N_2}\right)^2 Z_2 \tag{5-5}$$

Thus the impedance connected to the secondary side is referred to as the primary side by multiplying the impedance on the secondary side of the transformer by the square of the ratio of primary to secondary voltage which is also called *turn ratio*.*

Again makes use of Eqs. (5-3) and (5-4)

$$\dot{V}_1 \dot{I}_1 = \dot{V}_2 \dot{I}_2, \quad \dot{V}_1 \dot{I}_1^* = \dot{V}_2 \dot{I}_2^* \tag{5-7}$$

So the volt-amperes and complex power input to the primary winding equal the output of these same quantities from the secondary winding in an ideal transformer.

## Part 4 The Equivalent Circuit of a Practical Transformer

Now we study a practical transformer where (1) permeability is not infinite, (2) winding resistance is present, (3) losses occur in the iron core due to the cyclic changing of direction of the flux, and (4) not all the flux linking any one winding links the others.

When a sinusoidal voltage is applied to a transformer winding on an iron core with the secondary winding open a small current will flow in the primary such that in a well-designed transformer the maximum flux density $B_m$ occurs at the knee of the *B-H*, or saturation curve of the transformer.* This current is called the magnetizing current. Losses in the iron occur due, first, to the fact that the cyclic changes of the direction of the flux in the iron require energy which is dissipated as heat and called *hysteresis loss*. The second loss is due to the fact that circulating currents are induced in the iron due to the changing flux, and these currents produce an $|I|^2R$ loss in the iron called *eddy-current loss*. Hysteresis loss is reduced by the use of certain high grades of alloy steel for the core. Eddy-current loss is reduced by building up the core with laminated sheets of steel. With the second open the transformer's primary circuit is simply one of very high inductance due to the iron core. The current lags the applied voltage by slightly less than 90° and the component of current in phase with the voltage accounts for the energy loss in the core. In the equivalent circuit magnetizing current $I_E$ is taken into account by an inductive susceptance $B_L$ in parallel with a conductance G.

In the practical two-winding transformer some of the flux linking the primary winding does not link the secondary. This flux is proportional to the primary current and causes a voltage drop that is accounted for by an inductive reactance $x_1$, called *leakage reactance*, which is added in series with the primary winding of the ideal transformer. Similar leakage reactance $x_2$ must be added to the secondary winding to account for the voltage due to the flux linking the secondary but not the primary. When we also account for the resistances $r_1$ and $r_2$ of the windings we have the transformer model shown in Figure 5.1. In this model the ideal transformer is the link between the circuit parameters $r_1$, $x_1$, G, and $B_L$ added to the primary side of the transformer and $r_2$ and $x_2$ added to the secondary side.

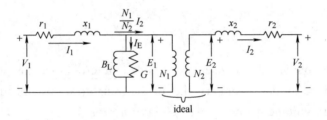

**Figure 5.1** The transformer equivalent circuit using the ideal transformer concept

The ideal transformer may be omitted in the equivalent circuit if we refer all quantities to either the high or the lowvoltage side of the transformer. For instance, if we refer all voltages, currents, and impedances of the circuit of Figure 5.1 to the primary circuit of the transformer having $N_1$ turns, and for simplicity let $a = N_1 / N_2$, we have the circuit of Figure 5.2. Very often we neglect magnetizing current because it is so small compared to the usual load currents. To further simplify the circuit we let

$$R_1 = r_1 + a^2 r_2 \text{ , and } X_1 = x_1 + a^2 x_2$$

to obtain the equivalent circuit of Figure 5.3. All impedances and voltages in the part of the circuit connected to the secondary terminals must now be referred to the primary side.

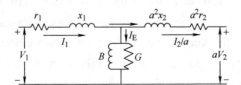 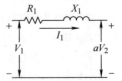

**Figure 5.2** the transformer equivalent circuit with path for magnetizing current

**Figure 5.3** the transformer equivalent circuit with magnetizing current neglected

# Chapter 5  Power Transformers

The following approximations can be made when the transformers are under load:

1. The voltage induced in a winding is directly proportional to the number of turns, the frequency, and the flux in the core.

2. The ampere-turns of the primary are equal and opposite to the ampere-turns of the secondary.

3. The apparent power input to the transformer is equal to the apparent power output.

4. The exciting current in the primary winding may be neglected.

## Part 5  Special Transformers

Many transformers are designed to meet specific industrial applications. Although they are special, they still possess the basic properties of the standard transformers.

**1. Autotransformer**

Consider a single transformer winding having $N_1$ turns, mounted on an iron core (Figure 5.4). The winding is connected to a fixed-voltage AC source $E_1$, and the resulting exciting current $I_0$ creates an AC flux $\Phi_m$ in the core. Suppose a tap C is taken off the winding so that there are $N_2$ turns between terminals A and C. Because the induced voltage between these terminals is proportional to the number of turns, $E_2$ is given by $E_2 = (N_2/N_1) \times E_1$.

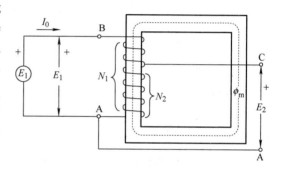

**Figure 5.4**  Autotransformer having $N_1$ turns on the primary and $N_2$ turns on the secondary

Clearly, this simple coil resembles a transformer having a primary voltage $E_1$, and a secondary voltage $E_2$. However, the primary terminals B-A and the secondary terminals C-A are no longer isolated from each other because of the common terminal A.

If we connect a load to secondary terminals C-A, the resulting current $I_2$ immediately causes a primary current $I_1$ to flow.

The B-C portion of the winding obviously carries current $I_1$. Therefore, according to Kirchhoff's current law, the C-A portion carries a current $(I_2-I_1)$. Furthermore, the MMF (Magnetic Motive Force) due to $I_1$, must be equal and opposite to the MMF produced by $(I_2-I_1)$. As a result, we have

$$I_1(N_1-N_2)=(I_2-I_1)N_2$$

which reduces to

$$I_1N_1=I_2N_2$$

Finally, assuming that both the transformer losses and exciting currents are negligible, the apparent power drawn by the load must be equal to the apparent power supplied by the source. Consequently,

$$E_1I_1=E_2I_2$$

The equations are identical to those of a standard transformer having a turn ratio $N_2/N_1$.

However, in this autotransformer, the secondary winding is actually part of the primary winding. In effect, an autotransformer eliminates the need for a separate secondary winding. As a result, autotransformers are always smaller, lighter, and cheaper than standard transformers of equal power output. The difference in size becomes particularly important when the ratio of transformation $E_1/E_2$ lies between 0.5 and 2. On the other hand, the absence of electrical isolation between the primary and secondary windings is a serious drawback in some applications.

Autotransformers are used to start induction motors, regulate the voltage of transmission lines, and, in general, transform voltages when the primary-to-secondary ratio is close to 1.

The great disadvantage of the autotransformer is that electrical isolation is lost, which is usually the decisive factor in favor of the ordinary connection in most applications. In power systems, three-phase autotransformers are frequently used to make small adjustments in bus voltages.

A conventional two-winding transformer can be changed into an autotransformer by connecting the primary and secondary windings in series. Depending upon how the connection is made, the secondary voltage may add to, or subtract from the primary voltage.* The basic operation and behavior of a transformer are unaffected by a mere change in external connection.

**2. Dual-voltage distribution transformer**

In the U.S., transformers that supply electric power to residential areas generally have two secondary windings, each rated at 120V. The windings are connected in series, and so the total voltage between the lines is 240V while that between the lines and the center tap is 120V (Figure 5.5). The center tap, called neutral, is always connected to the ground. The nominal rating of these distribution transformers ranges from 3kVA to 500kVA. They are mounted on poles to supply power to as many as 20 customers.

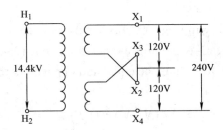

Figure 5.5  A distribution transformer with 120V/240V secondary

**3. High-impedance transformers**

The transformers we have studied so far are all designed to have a relatively low leakage

# Chapter 5  Power Transformers

reactance, ranging perhaps from 0.03 to 0.1 per unit. However, some industrial and commercial applications require much higher reactance, sometimes reaching values as high as 0.9 pu. Such high-impedance transformers are used in the following typical applications:

oil burners            arc welders              fluorescent lamps
EAF (electric arc furnaces)  neon signs        reactive power regulators

**4. High-frequency transformers**

In electronic power supplies, there is often a need to isolate the output from the input and to reduce the weight and cost of the unit. In other applications, such as in aircraft, there is a strong incentive to minimize weight. These objectives are best achieved by using a relatively high frequency compared to, say, 60Hz.* An increase in frequency reduces the size of such devices as transformers and inductors.

It is obvious that the increase in frequency has permitted a very large increase in the power capacity of the transformer. It follows that for a given power output a high-frequency transformer is much smaller, cheaper, more efficient, and lighter than a 60Hz transformer.

## NEW WORDS AND EXPRESSIONS

| 1.  | coil             | n.   | 线圈 |
| --- | ---------------- | ---- | ---- |
| 2.  | iron core        | n.   | 铁心 |
| 3.  | flux             | n.   | 流量,（磁）通量 |
| 4.  | ferromagnetic    | adj. | 铁磁的，铁磁体 |
| 5.  | confine          | vt.  | 限制，使局限于 |
| 6.  | winding          | n.   | 绕组，线圈 |
| 7.  | stack            | v.   | 堆叠 |
| 8.  | turn             | n.   | 一匝，盘绕的一圈 |
| 9.  | secondary        | adj. | 二级的，二次侧的，副的 |
| 10. | primary          | adj. | 初级的，原边的，一次侧的 |
| 11. | tertiary         | adj. | 第三的，第三位的 |
| 12. | autotransformer  | n.   | 自耦变压器 |
| 13. | mutual           | adj. | 相互的，共有的，互感的 |
| 14. | sinusoidally     | adv  | 按正弦（曲线）变化的 |
| 15. | permeability     | n.   | 磁导率，渗透性 |
| 16. | Faraday's law    |      | 法拉第定律 |
| 17. | instantaneous    | adj. | 瞬时的，即刻的 |
| 18. | phasor           | n.   | 相量 |
| 19. | turn ratio       | n.   | 变比，匝数比 |
| 20. | voltampere       | n.   | 伏安数，功率 |
| 21. | complex power    |      | 复数功率 |
| 22. | knee             | n.   | 弯曲部分；转弯处 |
| 23. | saturation       | n.   | 饱和 |
| 24. | hysteresis       | n.   | 磁滞（现象） |
| 25. | eddy-current     | n.   | 涡流 |
| 26. | laminated        | adj. | 薄板[片]状的，层积的，层压的 |

| | | | |
|---|---|---|---|
| 27. | lag | v. | 落后于,滞后 |
| 28. | susceptance | n. | 电纳(导纳的虚数分量) |
| 29. | leakage reactance | | 漏电抗,漏抗 |
| 30. | mount | vt. | 装配,设置,安放 |
| 31. | fixed-voltage | n. | 恒电压 |
| 32. | exciting current | | 励磁[激励]电流 |
| 33. | Kirchhoff's law | | 基尔霍夫定律 |
| 34. | apparent power | | 视在功率 |
| 35. | drawback | n. | 缺点,障碍 |
| 36. | induction motor | | 感应马达,感应电动机 |
| 37. | decisive | adj. | 决定性的 |
| 38. | dual | adj. | 双的,二重的,双重 |
| 39. | tap | n. | 分接头 |
| 40. | oil burner | | 燃油炉 |
| 41. | arc welder | | 电弧焊 |
| 42. | fluorescent lamp | | 荧光灯(管),日光灯(管) |
| 43. | neon sign | | 霓虹灯 |
| 44. | incentive | n. | 动机 |
| 45. | inductor | n. | 电感,感应器 |

# PHRASES

**1. as well as:** 也,又。注意: A <u>as well as</u> B 表示在 B 成立的同时 A 也成立。

Example: An autotransformer differs from the ordinary transformer in that the windings of the autotransformer are electrically connected <u>as well as</u> being coupled by a mutual flux, as will be described later.

自耦变压器与与普通变压器的差别在于,自耦变压器的绕组在经公共磁通耦合的同时在电气上是连接的,这在后文会有所描述。

**2. take into account:** 考虑

**in parallel with:** 与……并联,平行于

Example: In the equivalent circuit magnetizing current $I_E$ is <u>taken into account</u> by an inductive susceptance $B_L$ <u>in parallel with</u> a conductance $G$.

在等效电路中,励磁电流 $I_E$ 以感性电纳 $B_L$ 和电导并联 $G$ 的形式加以考虑。

**3. be proportional to:** 与……成正比

**account for:** 说明

**in series with:** 与……串联

Example: This flux <u>is proportional to</u> the primary current and causes a voltage drop that is <u>accounted for</u> by an inductive reactance $x_1$, called *leakage reactance*, which is added <u>in series with</u> the primary winding of the ideal transformer.

磁通与原边电流成正比,并且引起一个电压降。该电压降用一个与理想变压器的原

边绕组串联的称作漏抗的感抗来说明。

注意：在此例句中，a voltage drop 作为 cause 的宾语，而其本身所带的定语很长，在翻译时可以把定语从句与修饰的主体拆解开来，翻译起来会顺畅好多。

### 4. in favor of: 赞同，有利于

Example: The great disadvantage of the autotransformer is that electrical isolation is lost, which is usually the decisive factor in favor of the ordinary connection in most applications.

自耦变压器的主要缺点是丧失了电气隔离，而这通常是在大多数应用中对一般连接很有利的决定性因素。

## COMPLICATED SENTENCES

**1. An electric transformer is a device that changes AC electric energy at one voltage level into AC electric energy at another voltage level through the action of a magnetic field.**

译文：变压器是通过磁场作用将一个电压等级的电能转换为另一电压等级电能的设备。

说明：这是描述设备、装置的典型陈述句式。其大体结构是：设备名称 + 设备功能 + 实现途径。一般来说，其中的设备名称做主语，设备功能做谓语，实现途径做谓语。

**2. The first step-down of voltage from transmission levels to distribution levels is at the bulk-power substation, where the reduction is to a range of 34.5 to 138kV, depending, of course, upon transmission-line voltage.**

译文：从传输等级到配电等级的第一次降压发生在大容量变电站，降低到 34.5kV 至 138kV 的范围，当然，这取决于输电线的电压。

说明：of course 是插入语，在翻译时应提到所属的现在分词短语的最前面。the reduction is to 相当于 it is reduced to。

**3. In a power transformer the coils are placed on an iron core which is made of ferromagnetic material in order to confine the flux so that almost all of the flux linking any one coil links all the others.**

译文：在电力变压器中，线圈置于一个铁芯上。该铁心由铁磁材料制成，以约束磁通，使链接任一线圈的磁通几乎都通过其他线圈。

说明：在 the flux linking any one coil links all the others 从句中，the flux 是主语，links 是谓语，all the others 是宾语，linking any one coil 是修饰 the flux 的定语。

**4. The winding across which impedance or other load may be connected is called the secondary winding or output winding and any circuit elements connected to this winding are said to be on the secondary side of the transformer.**

译文：可以连接阻抗和其他负载的绕组被称为二次绕组或输出绕组，任何与其相连的电路元件都被称为在变压器的二次侧。

说明：这个长句由两个分句组成，由 and 联结。第一个分句中，across which impedance or other load may be connected 是一个介词置前的定语从句，修饰主语 winding。为了便于理解，这个句子可以拆解为：

Impedance or other load may be connected across a winding; the winding is called the secondary winding or output winding; Any circuit elements connected to this winding are said to be on the secondary side of the transformer.

**5. We shall begin our analysis by assuming that the flux varies sinusoidally in the core and that the transformer is ideal, which means that the permeability μ of the core is infinite and the resistance of the windings is zero.**

译文：开始分析前，我们先假定磁通在铁芯中按正弦规律变化，而且变压器是理想的，也就是说铁芯的磁导率 $\mu$ 无穷大而绕组的电阻为零。

说明：begin analysis by 意思为"以……开始分析"，为了句子通顺连贯，可以意译。which means that... 是定语从句中嵌套了宾语从句，可以解释为"这意味着，其意思是"，但如上面译文给出的说法似乎可以使句子更为通顺。

**6. One device to provide winding information is to place a dot at the end of each winding such that all dotted ends of windings are positive at the same time; that is, voltage drops from dotted to unmarked terminals of all windings are in phase.**

译文：提供绕组信息的一种方法是，在每个绕组的某一端放一个点，使所有绕组的带点一端同时为正；即，所有绕组从带点端到无标记端的电压降都同相。

说明：device 多用于表示"设备"，但是在本句中指的是"方法，策略"。such that 是 in such way that 的缩写，可以理解为"以这样一种方式"。

**7. Thus the impedance connected to the secondary side is referred to as the primary side by multiplying the impedance on the secondary side of the transformer by the square of the ratio of primary to secondary voltage which is also called *turn ratio*.**

译文：于是，将变压器二次侧的阻抗与一次侧和二次侧的电压比（也称为变比）的平方相乘，接在二次侧的阻抗可以折算到一次侧。

说明：词组 refer to 意思是"提到，查阅"，此处以被动语态的形式出现，可以理解为"折算到，折合为"。the square of 表示"……的平方"。句式 multiply A by B 意思是"A 与 B 相乘"。在 primary to secondary voltage 中省略了 primary 后面的 voltage，这种情况在英文科技文献中比较常见。

**8. When a sinusoidal voltage is applied to a transformer winding on an iron core with the secondary winding open a small current will flow in the primary such that in a well-designed transformer the maximum flux density $B_m$ occurs at the knee of the *B-H*, or saturation curve of the transformer.**

译文：当一个正弦电压施加到铁心上的一个变压器绕组而二次绕组开路时，会有一个小电流流过一次侧，从而在设计良好的变压器中，最大磁通密度 $B_m$ 出现在 B-H 曲线或叫饱和曲线的"膝点"（拐弯处）。

# Chapter 5  Power Transformers

说明：词组 be applied to 意思是"施加到，应用于"。Saturation curve 与 B-H curve 是同一事物的两种名称，因此在翻译时最好将其中间的连接词"or"译为"或者叫"而不是简单的译为"或"，以免使读者误以为这是两个不同的事物。为了便于理解，这个句子可以拆解为：

When a sinusoidal voltage is applied to a transformer winding on an iron core with the secondary winding open,（这里加了个逗号，把时间状语从句区隔开）a small current will flow in the primary, such that in a well-designed transformer the maximum flux density $B_m$ occurs at the knee of the *B-H* curve, also called saturation curve, of the transformer.

**9. Depending upon how the connection is made, the secondary voltage may add to, or subtract from the primary voltage.**

译文：基于连接方式，二次电压可以叠加到一次电压上，或者从一次电压中减掉。

说明：两个具有相同宾语的词组省掉了一个宾语，共用一个。这种情况在英文科技文献中比较常见。这个句子可以补全为：

Depending upon how the connection is made, the secondary voltage may add to the primary voltage or subtract from the primary voltage.

**10. These objectives are best achieved by using a relatively high frequency compared to, say, 60Hz.**

译文：通过采用相对（工频，比方说60Hz）较高的频率，这些目标得以很好的实现。

说明：插入到句中，前后以逗号分隔的词语 say，意思是"比方说"。另外，对于不太长的状语短语，在翻译时可以适当提前。

## ABBREVIATIONS (ABBR.)

1. MMF            magnetic motive force              磁动势
2. EAF            electric arc furnace              电弧炉

## SUMMARY OF GLOSSARY

1. 电力变压器
   unit transformer                （机端）单元变压器
   substation transformer          变电站变压器
   distribution transformer        配电变压器
   interconnecting transformer     联络变压器
2. 变压器绕组 windings
   primary winding                 一次绕组，原边绕组
   secondary winding               二次绕组，副边绕组
   tertiary winding                第三绕组
   input winding                   输入绕组
   output winding                  输出绕组
3. 变压器损耗 power loss in transformers
   hysteresis loss                 磁滞损耗

|  |  |  |
|---|---|---|
|  | eddy-current loss | 涡流损耗 |
| 4. | 变压器中的电流 currents in transformers |  |
|  | magnetizing current | 磁化电流，起磁电流 |
|  | circulating current | 环流 |
|  | eddy-current | 涡流 |
|  | exciting current | 励磁电流，激磁电流 |
| 5. | 电学定律 |  |
|  | Faraday's law | 法拉第定律 |
|  | Kirchhoff's current law | 基尔霍夫电流定律 |
| 6. | 电路元件的基本连接方式 |  |
|  | in series with | 串联 |
|  | in parallel with | 并联，旁路 |

# EXERCISES

**1. Translate the following words or expressions into Chinese.**

(1) iron core　　　　　(2) regulator　　　　　(3) secondary winding

(4) equivalent circuit　(5) cyclic　　　　　　(6) hysteresis loss

(7) eddy-current　　　(8) complex power　　(9) Kirchhoff's law

**2. Translate the following words or expressions into English.**

(1) 漏电抗　　　　　(2) 变比　　　　　　(3) 自耦变压器

(4) 饱和　　　　　　(5) 励磁电流　　　　(6) 视在功率

(7) 伏安数　　　　　(8) 电纳　　　　　　(9) 相量

**3. Fill in the blanks with proper words or expressions.**

(1) The great disadvantage of the autotransformer is that _____ is lost, which is usually the decisive factor in favor of the ordinary connection in most applications.

(2) An electric transformer is a device that changes AC electric energy at one voltage level into AC electric energy at another voltage level through the action of a _____.

(3) The winding across which impedance or other load may be connected is called the _____ winding or _____ winding and any circuit elements connected to this winding are said to be on the _____ side of the transformer.

(4) We shall begin our analysis by assuming that the flux varies sinusoidally in the core and that the transformer is ideal, which means that the permeability $\mu$ of the core is _____ and the resistance of the windings is _____.

(5) Thus the impedance connected to the secondary side is referred to as the primary side by multiplying the impedance on the secondary side of the transformer by the _____ of the ratio of primary to secondary voltage which is also called _____.

# Chapter 5　Power Transformers

# Word-Building (5)　auto-　自动，远程

## 1. auto-　表示：自动

| | | |
|---|---|---|
| automatic | *adj.* | 自动的，无意识的，机械的 |
| automotive | *adj.* | 自动（机，车，推进）的 |
| autosyn | *adj.* | 自动同步的 |
| automodulation | *n.* | 自动调制 |
| autoregulation | *n.* | 自动调整 |
| autocouple | *v.* | 自动耦合 |
| autoreclose | *v.* | 自动重合闸 |
| autotrack | *vt.* | 自动跟踪 |

## 2. auto-　表示：自治，自耦

| | | |
|---|---|---|
| autoexciting | *adj.* | 自激的 |
| autonomic | *adj.* | 自治的，自律的 |
| autooscillation | *n.* | 自振荡 |
| autotransformer | *n.* | 自耦变压器 |

# Chapter 6
# Distribution Systems and Loads

## Part 1  Introduction of Distribution Systems

Some industrial customers may be supplied at such voltage levels in a range of 34.5 to 138kV which belong to transmission levels, but most electric customers are supplied by feeders in the distribution system.

The primary distribution voltage ranges from 4kV to 34.5kV and is commonly between 11kV and 15kV. Small industrial customers are supplied by primary feeders at this voltage level from the primary system, which also supplies the distribution transformers providing secondary voltages over single-phase three-wire circuits for residential and commercial customers at 120/240V. * Other secondary circuits are three-phase four-wire systems.

As illustrated in Figure 6.1, the aggregated load represented at a transmission substation (Bus A) usually includes, in addition to the connected load devices, the effects of substation step-down transformers *, sub-transmission feeders, distribution feeders, distribution transformers, voltage regulators, and reactive power compensation devices.

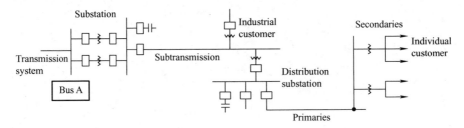

**Figure 6.1**  Power system configuration identifying parts of the system represented as load at a bulk power delivery point

## Part 2  Familiar Loads and Their Classification

Roughly, Loads are devices that consume electric energy or electric power consumed by customers mainly in the distribution system. * The Loads absorb electric energy from the power system and convert it into energy of other forms, which may be heat energy to heat water or to melt irons, or mechanical energy to drive a machine, etc. *

They can be divided into various classes according to different conditions.

# Chapter 6  Distribution Systems and Loads

For instance, the three fundamental load classes, according to the situations where they are used, are industrial, commercial, and residential loads.

The classification by service requirements is also used sometimes. From this viewpoint, some loads require simply enough electric energy, in other words, the quantity is the primary desire. This kind of load, especially those that consume energy calculated by standard coals of more than 10 thousand tons per year, has a name of high energy consuming loads.* A metallurgical plant with one or more large capacity EAFs (Electric Arc Furnace) for melting metal is the typical instance. For some other industries, high power supply quality is the primary desire. For example, banks and settlement centers require strictly continuous service, which is of extreme importance for them to keep the data intact and consistent. As to power supply for agricultural irrigation, almost no special requirement exists.

The load status is very important for a power system. The power system is called a *large mode* of operation under heavy load conditions, and similarly, the term *small mode* for light loads.

Table 6.1 lists some familiar loads.

Table 6.1  Familiar Loads

| Load class | | Industrial loads | Commercial loads | Residential loads |
|---|---|---|---|---|
| Rotating | Motors | Induction motors<br>Fan motors | Fan motors | Dishwasher<br>Clothes washer<br>Clothes dryer |
| | Compressors | | Air conditioner | Refrigerator |
| Heating | | Arc furnace<br>Converter | | Water heaters<br>Oven<br>Deep fryer |
| Lighting | | Incandescent lights<br>Fluorescent lights | Incandescent lights<br>Fluorescent lights | Incandescent lights<br>Fluorescent lights |
| Electronic | | Converters | | Television |

# Part 3  Loads Models for Power System Study

A typical load bus is composed of a large number of devices such as fluorescent and incandescent lamps, refrigerators, heaters, compressors, motors, furnaces, and so on. The exact composition of the load is difficult to estimate. Also, the composition changes depending on many factors including time (hour, day, and season), weather conditions, and the state of the economy.

Even if the load composition were known exactly, it would be impractical to represent each individual component as there are usually millions of such components in the total load supplied by a power system. Therefore, load representation in system studies is usually based on a considerable amount of simplification.

The load models are traditionally classified into two broad categories: *static models* and *dynamic models*.

**1. Static Loads**

A static load model expresses the characteristic of the load at any instant of time as algebraic functions of the bus voltage magnitude and frequency at that instant.*

Traditionally, the voltage dependency of load characteristics has been represented by the exponential model:

$$P = P_0(V/V_0)^a , \quad Q = Q_0(V/V_0)^b$$

where P and Q are active and reactive parts of the load when the bus voltage magnitude is *V*. The subscript $_0$ identifies the values of the respective variables at the initial operating condition. With the exponents a and b equal to 0, 1, or 2, the model represents constant power, constant current, or constant impedance characteristics, respectively. For composite loads, their values depend on the aggregate characteristics of load components.

An alternative model which has been widely used to represent the voltage dependency of loads is the polynomial model:

$$P = P_0[p_1(V/V_0)^2 + p_2(V/V_0) + p_3] , \quad Q = Q_0[q_1(V/V_0)^2 + q_2(V/V_0) + q_3]$$

This model is commonly referred to as the ZIP model, as it is composed of constant impedance (Z), constant current (I), and constant power (P) components.

**2. Dynamic Loads**

There are, however, many cases where it is necessary to account for the dynamics of load components.* Typically, motors consume 60% to 70% of the total energy supplied by a power system. Therefore, the dynamics attributable to motors are usually the most significant aspects of the dynamic characteristics of system loads.

Other dynamic aspects of load components that require consideration in stability studies include the following:

(1) Extinction of discharge lamps below a certain voltage and their restart when the voltage recovers. Discharge lamps include mercury vapor, sodium vapor, and fluorescent lamps. These extinguish at a voltage in the range of 0.7 to 0.8 pu. When the voltage recovers, the restart after a 1 or 2-second delay.

(2) Operation of protective relays, such as thermal and overcurrent relays. Many industrial motors have starters with electromagnetically held contactors. These drop open at voltages in the range of 0.55 to 0.75 pu; the dropout time is on the order of a few cycles. Small motors on refrigerators and air conditioners have only thermal overload protections, which typically trip in about 10 to 30 seconds.

(3) Thermostatic control of loads, such as space heaters/coolers, water heaters, and refrigerators. Such loads operate longer during low-voltage conditions. As a result, the total number of these devices connected to the system will increase in a few minutes after a volt-

# Chapter 6  Distribution Systems and Loads

age drop. Air conditioners and refrigerators also exhibit such characteristics under sustained low-frequency conditions.

(4) Response of ULTCs (Under-Load Tap Changing) on distribution transformers, voltage regulators, and voltage-controlled capacitor banks. These devices are not explicitly modeled in many studies. In such cases, their effects must be implicitly included in the equivalent load that is represented at the bulk power delivery point.

As these devices restore distribution voltages following a disturbance, the power supplied to voltage-sensitive loads is restored to the pre-disturbance levels.

## Part 4  Acquisition of Load Model Parameters

There are two basic approaches to the determination of system-load characteristics: *measurement-based approach* and *component-based approach*.

### 1. Measurement-Based Approach

In this approach, the load characteristics are measured at representative substations and feeders at selected times of the day and season. These are used to extrapolate the parameters of loads throughout the system. The load models can be derived from staged tests or actual system transients. An alternative approach is that in which load characteristics are monitored continuously from naturally occurring system variations.

*Steady-state load-voltage characteristics.* Composite load characteristics are normally measured at the highest voltage level for which the voltage of the radially connected load can be adjusted by transformer tap changers. This is usually the highest distribution voltage level. The steady-state characteristics can be determined by adjusting the load voltage via the transformer tap changers over a range of voltage above and below the nominal value.* Distribution tap changers and switched capacitors must be blocked to obtain meaningful results. The measured responses of voltage, active power, and reactive power are fitted to polynomial and / or exponential expressions.

*Steady-state load-frequency characteristics.* It is usually much more difficult to measure system load-frequency characteristics. To measure composite load-frequency characteristics, an isolated system must be formed and the frequency must vary over the desired range. To obtain valid data, care must be taken to separate the effects of voltage changes and frequency changes. From many of the results reported in the literature, this is not done, and the $dP/df$ characteristics determined is a composite of the effects due to frequency changes and the resulting voltage changes.

*Dynamic load-voltage characteristics.* The small-signal dynamic characteristics of composite

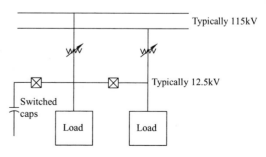

**Figure 6.2**  Typical station configuration for testing load characteristics

loads can be determined relatively easily from simple system tests. Figure 6.2 shows the testing configuration that may be employed when loads are supplied by two tap-changing transformers.

Initially, one tap changer is adjusted upward and the other downward by a few taps, keeping the load voltage constant. One of the transformers is then tripped; this produces not only a voltage magnitude change but also an instantaneous angle change at the load bus. By varying the initial tap positions, it is possible to obtain a range of voltage changes in both positive and negative directions. By selecting the tap positions appropriately, it is also possible to produce an angle change with only a very small voltage change. This is useful in separating the effects of voltage magnitude and angle change.

If there is a switched capacitor bank at the load bus, it can be switched in and out to produce a voltage magnitude change at the load without an angle change.

The responses obtained in this fashion are essentially small-signal.

A least square difference technique is used to optimize the match between the measured and model response as the model parameters are adjusted. The result is a model whose parameters are chosen to match the small-signal response. However, because of the realistic model structure chosen, the model is likely to give reasonable results under large-disturbance conditions.

**2. Component-Based Approach**

This approach was developed by EPRI (Electric Power Research Institute) under several research projects beginning in 1976. It involves building up the load model from information on its constituent parts as illustrated in Figure 6.3. The load supplied at a bulk power delivery point is categorized into load classes such as residential, commercial, industrial, agricultural, and mining. Each category of load class is represented in terms of load components such as lighting, air conditioning, space heating, water heating, and refrigeration.

The characteristics of individual appliances have been studied in detail and techniques have been developed to aggregate individual loads to produce a composite load model.

The EPRI LOADSYN program converts data on the load class mix, components, and their characteristics into the form required for power flow and stability programs. Typical default data have been developed for load composition and characteristics for each class of load.

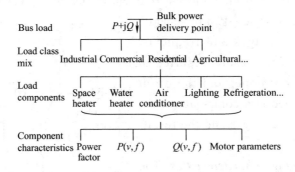

**Figure 6.3** Component-based modeling approach

# Chapter 6  Distribution Systems and Loads

## NEW WORDS AND EXPRESSIONS

| | | | |
|---|---|---|---|
| 1. | feeder | n. | 馈电线 |
| 2. | aggregated | adj. | 合计的，集合的 |
| 3. | delivery point | | 交收点 |
| 4. | standard coal | n. | 标准煤 |
| 5. | high energy consuming | | 高耗能的 |
| 6. | metallurgical | adj. | 冶金的 |
| 7. | settlement center | | 结算中心 |
| 8. | intact | adj. | 完整无缺的 |
| 9. | agricultural irrigation | | 农业灌溉 |
| 10. | compressor | n. | 压缩机 |
| 11. | refrigerator | n. | 电冰箱，冷藏库 |
| 12. | oven | n. | 烤箱，烤炉，灶 |
| 13. | deep fryer | | 油炸锅，煎锅 |
| 14. | incandescent light | | 白炽灯 |
| 15. | fluorescent light | | 荧光灯 |
| 16. | converter | n. | 转炉；变流器 |
| 17. | dynamic | adj. | 动态的 |
| 18. | algebraic | adj. | 代数的 |
| 19. | exponential | adj. | 指数的，幂数的 |
| 20. | exponent | n. | 指数 |
| 21. | subscript | n. | 下标 |
| 22. | polynomial | adj. | 多项式的 |
| 23. | attributable | adj. | 可归于…的 |
| 24. | discharge lamp | | 放电管 |
| 25. | mercury | n. | 水银，汞 |
| 26. | vapour | n. | 蒸汽 |
| 27. | sodium | n. | 钠 |
| 28. | overcurrent | n. | 过电流 |
| 29. | electromagnetically | adv | 电磁（控制）地 |
| 30. | contactor | n. | 电流接触器 |
| 31. | dropout | n. | 退出 |
| 32. | trip | v. | 松开，跳闸 |
| 33. | thermostatic | adj. | 温度调节装置的，恒温的 |
| 34. | sustained | adj. | 持续不变的，相同的 |
| 35. | capacitor bank | | 电容器组（合） |
| 36. | explicitly | adv | 明白地，明确地 |
| 37. | implicitly | adv | 含蓄地，暗中地 |
| 38. | pre-disturbance | n. | 扰动前 |
| 39. | acquisition | n. | 获得；采集 |
| 40. | extrapolate | v. | 推断 |
| 41. | staged test | | 安排和进行的测试 |
| 42. | literature | n. | 文献 |
| 43. | negative direction | | 逆向，反方向 |

| | | | |
|---|---|---|---|
| 44. | least square | n. | 最小二乘 |
| 45. | optimize | vt. | 使最优化 |
| 46. | constituent | adj. | 组成的 |
| 47. | mining | n. | 采矿，矿业 |
| 48. | LOADSYN | | 一个仿真软件名称 |
| 49. | default | n. | 默认（值），缺省（值） |
| 50. | power factor | | 功率因数 |

## PHRASES

**1. in other words: 换句话说，换言之**

Example: From this viewpoint, some loads requires simply enough electric energy, in other words, the quantity is the primary desire.

从这个观点出发，一些负荷只是简单地要求充足的电能，换句话说，数量是第一位的要求。

**2. be of extreme importance for: 对…具有极其重要的意义，对…极为重要**

Example: For example, banks and settlement centers require strictly continuously service, which is of extreme importance for them to keep the data intact and consistent.

例如，银行和结算中心严格要求持续供电，这对其保持数据的完整性和一致性及其重要。

**3. attributable to: 可归于……，归因于……**

Example: Therefore, the dynamics attributable to motors are usually the most significant aspects of dynamic characteristics of system loads.

因此，与电动机有关的动态通常是负荷动特性当中最重要的方面。

**4. on the order of: 属于…一类的，与…相似的。在此表示：…的数量级**

Example: the dropout time is on the order of a few cycles.
退出时间是几个周期的量级。

## COMPLICATED SENTENCES

**1. Small industrial customers are supplied by primary feeders at this voltage level from the primary system, which also supplies the distribution transformers providing secondary voltages over single-phase three-wire circuits for residential and commercial customers at 120/240V.**

译文：小型工业用户由一次系统中此电压等级的一次馈线供电。一次系统也经过单相三线电路为居民用电和商业用户提供 120/240V 二次电压的配电变压器供电。

说明：这个句子带有一个由 which 引导的很长的定语从句，修饰的主体是 primary system。在定语从句中，providing 及后面的内容构成一个现在分词短语，修饰 distribu-

# Chapter 6　Distribution Systems and Loads

tion transformers。词组 provide...for... 表示："为…提供…"。为了便于理解，该句子可以拆解为：

Small industrial customers are supplied by primary feeders at this voltage level from the primary system; the primary system also supplies the distribution transformers; the distribution transformers provide secondary voltages for residential and commercial customers at 120/240V over single-phase three-wire circuits.

**2. As illustrated in Figure 6.1, the aggregated load represented at a transmission substation (Bus A) usually includes, in addition to the connected load devices, the effects of substation step-down transformers....**

译文：如图 6.1 所示，在输电变电站（母线 A）所给出的集合负荷中，除了所连接的负荷设备以外，通常包括变电站降压变压器的影响……。

说明：in addition to 表示"除了……之外"，它与后面所带宾语构成状语短语，作为插入语放在了句子中间，在翻译时应提到整个句子的谓语之前。

**3. Roughly, Loads are devices that consume electric energy or electric power consumed by customers mainly in the distribution system.**

译文：概略地讲，负荷主要是指在配电网中消耗电能的设备或者由用户消耗的功率。

说明：这个句子不长，但是翻译的时候很容易出错。这里给出了 load 在两个方面的含义：一是 device，一是 electric power，这两个概念都主要体现在配电网中。为了便于理解，这个句子可以拆解为：

Roughly, Loads are devices that consume electric energy mainly in the distribution system; or loads are electric power consumed by customers mainly in distribution system.

**4. The Loads absorb electric energy from the power system and convert it into energy of other forms, which may be heat energy to heat water or to melt irons, or mechanical energy to drive a machine, etc.**

译文：负荷从电力系统中吸收电能，并将其转换为其他形式的能量，可能是烧水或熔化钢铁的热能，或者是驱动机器的机械能，等等。

说明：由 which 引导的定语从句带有两个由 or（第二个）连接的并列宾语，分别是 heat energy 和 mechanical energy，这两个宾语又都各自带有定语。which 引导的定语从句修饰的主体是 energy of other forms。为了便于理解，这个句子可以拆解为：

The Loads absorb electric energy from the power system and convert it into energy of other forms; energy of other form may be heat energy; heat energy heats water or to melts irons; energy of other form may be mechanical energy; mechanical energy drive a machine; etc.

**5. This kind of load, especially those that consume energy calculated by standard coals of more than 10 thousand tons per year, has a name of high energy consuming loads.**

译文：这种负荷，尤其是那些按标准煤计算每年消耗能量超过一万吨（标准煤）的负荷，有一个名字叫高耗能负荷。或者：这种负荷，尤其是那些年耗能万吨标准煤以上的负荷，被称为高耗能负荷。

**6. A static load model expresses the characteristic of the load at any instant of time as algebraic functions of the bus voltage magnitude and frequency at that instant.**

译文：静态负荷模型作为任一（特定的）时刻母线电压大小和频率的代数函数，表征该时刻的负荷特性。

说明：句中先后出现了 at any instant of time 和 at that instant，所指内容是相同的，因此在翻译时可以根据在译句中出现的顺序，适当调整。一般先出现的，可翻译为"任一时刻"，后出现的译为"该时刻"。这种情况在英文文献中比较常见，可以依此类推。

**7. There are, however, many cases where it is necessary to account for the dynamics of load components.**

译文：然而，有很多实例应该考虑负载元件的动态性。

说明：词组 account for 表示"说明，考虑"。However 虽然出现在句子中间，但在翻译时应提到句首。

**8. The steady-state characteristics can be determined by adjusting the load voltage via the transformer tap changers over a range of voltage above and below the nominal value.**

译文：稳态特性可以通过用变压器分接头变换器将负荷电压调节到标称值上下的一个范围来确定。

说明：句子出现了一连串的介词，在翻译时一定要注意各自的作用范围，还要注意整个译句的顺畅达意。

## ABBREVIATIONS (ABBR.)

1. EAF　　　　electric arc furnace　　　　电弧炉
2. ULTC　　　under-load tap changing　　有载调节分接头
3. EPRI　　　 electric power research institute　　电力研究院

## SUMMARY OF GLOSSARY

1. 负荷分类 load classes
   industrial loads　　　　　　　　　　工业负荷
   commercial loads　　　　　　　　　商业负荷
   residential loads　　　　　　　　　　居民负荷
   agricultural loads　　　　　　　　　农业负荷
   mining loads　　　　　　　　　　　矿业负荷
2. 馈电线路 feeders
   primary feeders　　　　　　　　　　一次馈线
   sub-transmission feeders　　　　　　中高压馈线
   distribution feeders　　　　　　　　配电馈线

# Chapter 6　Distribution Systems and Loads

3. 供电线路形式
    single-phase three-wire　　　　　　　　单相三线（制）
    three-phase four-wire　　　　　　　　　三相四线（制）
4. 系统运行方式
    large mode　　　　　　　　　　　　　大方式
    small mode　　　　　　　　　　　　　小方式
5. 负荷状况 load status
    heavy loads　　　　　　　　　　　　　重负荷
    light loads　　　　　　　　　　　　　轻负荷
6. 功率构成 power
    apparent power　　　　　　　　　　　视在功率
    active power　　　　　　　　　　　　有功功率
    reactive power　　　　　　　　　　　无功功率
7. 静态负荷特性 static load characteristics
    constant power　　　　　　　　　　　恒功率
    constant current　　　　　　　　　　恒电流
    constant impedance　　　　　　　　　恒阻抗
8. 静态负荷模型 static load model
    exponential model　　　　　　　　　指数模型
    polynomial model　　　　　　　　　多项式模型
9. 负荷建模方法 approaches to determine load characteristics
    measurement-based approach　　　　测辨法
    component-based approach　　　　　合成法

# EXERCISES

**1. Translate the following words or expressions into Chinese.**

(1) configuration　　　(2) compressors　　　(3) category
(4) contactor　　　　　(5) trip　　　　　　　(6) restore
(7) approach　　　　　(8) acquisition　　　　(9) monitor

**2. Translate the following words or expressions into English.**

(1) 馈电线　　　　　　(2) 电容器组　　　　　(3) 扰动前
(4) 最小二乘　　　　　(5) 高耗能的　　　　　(6) 功率因数
(7) 指数　　　　　　　(8) 变量　　　　　　　(9) 暂态

**3. Fill in the blanks with proper words or expressions.**

(1) The three fundamental load classes, according to the situations where they are used, are _____, _____, and _____ loads.

(2) The Loads absorb _____ energy from the power system and convert it into energy of other forms, which may be _____ energy to heat water or to melt irons, or _____ energy to drive a machine, etc.

(3) Therefore, the dynamics attributable to _____ are usually the most significant aspects of the dynamic characteristics of system loads.

(4) There are two basic approaches to the determination of system-load characteristics: _____ approach and _____ approach.

## Word-Building (6)　re-; im-　重，再；不，非

### 1. re- 前缀，表示：再次…，重…

| | | | | | |
|---|---|---|---|---|---|
| arrange | 安排 | rearrange | v. | 重安排，再设置 |
| close | 合闸 | reclose | v. | 重合闸 |
| cover | 保护，覆盖 | recover | v. | 复原，恢复 |
| establish | 建立 | reestablish | v. | 重建，恢复 |
| gain | 达到，获得 | regain | v. | 重新达到，恢复 |
| heater | 加热器 | reheater | n. | 再热器 |
| lay | v. 放置 | relay | n. | 继电器 |
| name | 命名 | rename | vt. | 重命名 |
| new | adj. 新，新的 | renew | v. | 更新 |
| produce | 出示，生产 | reproduce | v. | 再生，复制，使重现 |
| route | n. 路线　v. 发送 | reroute | v. | 改线，改道发送 |
| set | 设置 | reset | v. | 重新设置 |
| start | 开始，启动 | restart | v. | 重新开始，重新启动 |
| store | 存储 | restore | v. | 恢复，还原 |
| try | 试 | retry | v. | 重试 |

### 2. im- 前缀，表示"否定，与…相反"之义（用于 b, m, p 之前），可译为：不，非

| | | | | |
|---|---|---|---|---|
| balance | 平衡，均衡 | imbalance | n. | 不平衡，不均衡 |
| becilic | n. 聪明 | imbecilic | adj. | 不聪明的，低能的，愚笨的 |
| maculate | 有斑点的 | immaculate | adj. | 没有斑点的，完美的 |
| material | 物质的；重要的 | immaterial | adj. | 非物质的；不重要的，非本质的 |
| mature | 成熟的 | immature | adj. | 不成熟的，未完全发展的 |
| palpably | 能感知地 | impalpably | adv. | 不能感知地，难以理解地 |
| parity | 相等，均衡 | imparity | n. | 不平等，不同，不匀称 |
| partial | 偏袒的 | impartial | adj. | 不偏不倚的，公平的 |
| possible | 可能的 | impossible | adj. | 不可能的，不会发生的 |
| practical | 实际的 | impractical | adj. | 不切实际的 |

# Unit 2

# Review

## I. Summary of Glossaries: Power Delivery and Consumption

1. 输电线，电力线 **power line**
   - transmission line —— 输电线，传输线
   - overhead line —— 架空线
   - cable —— 电缆
   - feeder —— 馈电线

2. 电路元件的基本连接方式
   - series- in series with —— 串联
   - shunt - in parallel with —— 并联，旁路

3. 电压等级变化
   - step up —— 升压
   - step down —— 降压

4. 电力变压器 **power transformer**
   - unit transformer —— （机端）单元变压器
   - substation transformer —— 变电站变压器
   - distribution transformer —— 配电变压器
   - interconnecting transformer —— 联络变压器

5. 变压器绕组 **windings**
   - primary winding —— 一次绕组，原边绕组
   - secondary winding —— 二次绕组，副边绕组
   - tertiary winding —— 第三绕组

6. 变压器中的电流和损耗 **current and power loss in transformers**
   - magnetizing current —— 磁化电流，起磁电流
   - exciting current —— 励磁电流，激磁电流
   - hysteresis loss —— 磁滞损耗
   - eddy-current loss —— 涡流损耗

7. 主要负荷分类 **load classes**
   - industrial loads —— 工业负荷
   - commercial loads —— 商业负荷
   - residential loads —— 居民负荷

8. 功率构成 power
   - apparent power 视在功率
   - active power 有功功率
   - reactive power 无功功率

## II. Abbreviations (Abbr.)

| | | | |
|---|---|---|---|
| 1. | ULTC | under-load tap changing | 有载调节分接头 |
| 2. | EPRI | electric power research institute | 电力研究院 |
| 3. | EHV | extra-high voltage | 超高压 |
| 4. | MMF | magnetic motive force | 磁动势 |

# Special Topic (2)　Passive Elements and Impedance

Basic passive elements and relative parameters are shown in Table 1. The impedance and admittance and their units are shown in Table 2.

Table 1　Passive circuit elements and their parameters

| 无源元件及其电气参数 | | | | 电气量的量纲 | | |
|---|---|---|---|---|---|---|
| Symbol | Element | Quantity | Meaning | Symbol | Name | Meaning |
| $R$ | Resistor | Resistance | 电阻 | Ω | Ohm | 欧姆 |
| $L$ | Inductor | Inductance | 电感 | H | Henry | 亨利 |
| $C$ | Capacitor | Capacitance | 电容 | F | farad | 法拉 |
| $M$ | | Mutual Inductance | 互感 | H | Henry | 亨利 |
| $X$ | Reactor | Reactance | 电抗 | Ω | Ohm | 欧姆 |
| $G$ | Conductor | Conductance | 电导 | S | siemens | 西门子 |

Table 2　Impedance, admittance and their units

| 阻抗参数 | | | | 电气量的量纲 | | |
|---|---|---|---|---|---|---|
| $R$ | | Resistance | 电阻 | Ω | Ohm | 欧姆 |
| $X$ | | Reactance | 电抗 | | | |
| $Z$ | $Z=R+jX$ | Impedance | 阻抗 | | | |
| $G$ | | Conductance | 电导 | S | siemens | 西门子 |
| $B$ | | Susceptance | 电纳 | | | |
| $Y$ | $Y=G+jB$ | Admittance | 导纳 | | | |

# Reading Material —— Per-Unit Value

Power transmission lines are operated at voltage levels where the kilovolt is the most convenient unit to express voltage. Because of the large amount of power transmitted kilowatts or megawatts and kilovoltamperes or megavoltamperes are the common terms. However, these quantities as well as amperes and ohms are often expressed as a percent or per unit of a base or reference value specified for each. For instance, if a base voltage of 120kV is chosen, voltages of 108,120, and 126kV become 0.9, 1.00, and 1.05 per unit (p.u.), or 90%, 100%, and 105%, respectively. The per-unit value of any quantity is defined as the ratio of the quantity to its base value expressed as a decimal. The ratio in percent is 100 times the value per unit. Both the percent and per-unit methods of calculation are simpler than the use of actual amperes, ohms, and volts. The per-unit method has an advantage over the percent method because the product of two quantities expressed in per unit is expressed in per unit itself, but the product of two quantities expressed in percent must be divided by 100 to obtain the result in percent.*

Voltage, current, kilovoltamperes, and impedance are so related that the selection of base values for any two of them determines the base values of the remaining two.* If we specify the base values of current and voltage, base impedance and base kilovolt amperes can be determined. The base impedance is that impedance which will have a voltage drop across it equal to the base voltage when the current flowing in the impedance is equal to the base value of the current.* The base kilovolt amperes in single-phase systems is the product of base voltage in kilovolts and base current in amperes. Usually, base megavolt amperes and base voltage in kilovolts are the quantities selected to specify the base. For single-phase systems, or three-phase systems where the term current refers to line current, where the term voltage refers to voltage to neutral, and where the term kilovoltamperes refers to kilovoltamperes per phase, the following formulas relate the various quantities:

$$\text{Base current(A)} = \frac{\text{base capacity}_{1\Phi}(\text{kVA})}{\text{base voltage}_{LN}(\text{kV})} \tag{1}$$

$$\text{Base impedance}(\Omega) = \frac{\text{base voltage}_{LN}(\text{V})}{\text{base current(A)}} \tag{2}$$

$$\text{Base impedance}(\Omega) = \frac{(\text{base voltage}_{LN})^2 \times 1000(\text{kV})}{\text{base}_{1\Phi}(\text{kVA})} \tag{3}$$

$$\text{Base impedance}(\Omega) = \frac{(\text{base voltage}_{LN})^2(\text{kV})}{\text{base}_{1\Phi}(\text{MVA})} \tag{4}$$

$$\text{Base power}_{1\Phi}(\text{kW}) = \text{base}_{1\Phi}(\text{kVA}) \tag{5}$$

$$\text{Base power}_{1\Phi}(\text{MW}) = \text{base}_{1\Phi}(\text{MVA}) \tag{6}$$

$$\text{Per-unit impedance of a circuit element} = \frac{\text{actual impedance}(\Omega)}{\text{Base impedance}(\Omega)} \tag{7}$$

In these equations, the subscripts 1Φ and LN denote "per phase" and "line-to-neutral", respectively, where the equations apply to three-phase circuits. If the equations are used for a single-phase circuit, base voltage$_{LN}$ means the voltage across the single-phase line, or line-to-ground voltage if one side is grounded.

Since three-phase circuits are solved as a single line with a neutral return, the bases for quantities in the impedance diagram are kilovolt amperes per phase, and kilovolts from line to neutral. Data are usually given as total three-phase kilovolt amperes or megavolt amperes and line-to-line kilovolts. Because of this custom of specifying line-to-line voltage and total kilovolt amperes or megavolt amperes, confusion may arise regarding the relation between the per-unit value of line voltage and the per-unit value of phase voltage. Although a line voltage may be specified as a base, the voltage in the single-phase circuit required for the solution is still voltage-neutral. The base voltage to neutral is the base voltage from line to line divided by 1.732. Since this is also the ratio between line-to-line and line-to-neutral voltages of a balanced three-phase system, the per-unit value of a line-to-neutral voltage on the line-to-neutral voltage base is equal to the per-unit value of the line-to-line voltage at the same point on the line-to-line voltage base if the system is balanced.* Similarly, the three-phase kilovolt amperes is three times the kilovolt amperes per phase, and the three-phase kilovoltamperes base is three times the base kilovolt amperes per phase. Therefore, the per-unit value of the three-phase kilovoltamperes on the three-phase kilovolt ampere base is identical to the per-unit value of the kilovolt amperes per phase on the kilovolt ampere-per-phase base.

## New Words and Expressions

| | | | |
|---|---|---|---|
| 1. | per-unit value | n. | 标幺值 |
| 2. | formula | n. | 公式 |
| 3. | neutral return | n. | 中性线回路 |

## Summary of Glossaries

| | |
|---|---|
| per-unit value (p.u.) | 标幺值 |
| base value | 基值 |
| reference value | 参考值 |
| actual value | 实际值，有名值 |
| percent value | 百分值，百分数 |

## Notes

**1. The per-unit method <u>has an advantage over</u> the percent method because <u>the product of</u> two quantities expressed in per unit is expressed in per unit itself, but the product of two quantities expressed in percent must be divided by 100 to obtain the result in percent.**

译文：标幺值法胜过百分数法，这是因为，两个用标幺值表示的量的乘积直接就是标幺值，而两个用百分数表示的量的乘积还必须除以 100 才能得到百分数形式的结果。

说明：词组 have an advantage over 表示"优于，胜过"。the product of 表示"……的乘积"。

**2. Voltage, current, kilovoltamperes, and impedance are <u>so</u> related <u>that</u> selection of base values for any two of them determines the base values of the remaining two.**

译文：电压、电流、千伏安和阻抗是相关联的，因而选择其中任意两个作为基值也决定了其余两个的基值。

说明：词组 so that 表示"如此……以至，因而"。

**3. The base impedance is that the impedance which will have a voltage drop across it <u>equal to</u> the base voltage when the current flowing in the impedance <u>is equal to</u> the base value of the current.**

译文：基值阻抗是那样的阻抗，当流过该阻抗的电流等于电流的基值时，该阻抗上的压降等于电压的基值。

说明：词组 equal to 和 be equal to 的意思是一样的，都表示"等于……"，前者 equal 是动词，而后者 equal 是形容词。该句子的宾语 that impedance 带有很长的定语，如果直接作为定语翻译，句子比较拗口。为此，可以做适当的语序调整。为了便于理解，这个句子可以拆解为：

The base impedance is that impedance; when the current flowing in the impedance is equal to the base value of the current, the impedance will have a voltage drop across it equal to the base voltage.

**4. Since this is also the ratio between line-to-line and line-to-neutral voltages of a balanced three-phase system, the per-unit value of a line-to-neutral voltage on the line-to-neutral voltage base is equal to the per-unit value of the line-to-line voltage at the same point on the line-to-line voltage base if the system is balanced.**

译文：既然这也是平衡的三相系统中的线电压（线对线的电压）和相电压（线对中性点的电压）之比，那么如果该系统是平衡的，基于相电压基值的相电压标幺值就等于同一点上基于线电压基值的线电压标幺值。

说明：句子很长，但是结构不算太复杂。为了便于理解和翻译，该句子可以拆解为：

Since this is also the ratio between line-to-line and line-to-neutral voltages of a balanced three-phase system, then if the system is balanced, the per-unit value of a line-to-neutral voltage on the line-to-neutral voltage base is equal to the per-unit value of the line-to-line voltage at the same point on the line-to-line voltage base.

# Knowledge & Skills (2)　Inversion and Intensive Mood

### 1. 倒装句（Inversion）

倒装句，是为了强调某一句子成分而改变其结构顺序的句子。倒装是一种修辞方式，用颠倒词句的次序来达到加强语势、语调和突出语意等效果。

（1）当主语较长而谓语又很短时，为避免头重脚轻，句子往往用主谓倒装的形式来表达。

例1：In 1953 came the first 345kV line.

1953年出现了第一条345kV线路。或：1953年第一条345kV线路出现了。

说明：这是一个倒装句，主语是the first 345kV line。这个句子可以按习惯顺序写为The first 345kV line came in 1953.

（2）当连续描述类似的事情时，为了避免重复，可以用do(es) so来代替类似的内容，并写为so do(se)引导的倒装形式。

例1：The distribution of breakdowns for a given gap follows with some exceptions approximately normal or Gaussian distribution, as does the distribution of overvoltages on the system.

对于给定的气隙而言，击穿的分布除了个别例外近似遵循正态分布或高斯分布，系统中的过电压分布也是这样。

说明：逗号后面的分句是一个倒装句，以as does为特征构成句子的倒装。这个句子相当于

... the distribution of overvoltages on the system does so.

**2. 强调语气（Intensive Mood）**

有时为了加强语气，在句中加入动词do(es)，它不改变句子结构和基本内容，只起强调作用，意思相当于副词indeed"确实"。

例1：However, insulation does conduct some current and so must be regarded as a material of very high resistivity.

然而，绝缘部分确实传导一些电流，因此必须被认为是一种具有很高电阻率的材料。

说明：句中does表示强调，可译为"确实"。and连接的两个谓语动词does和must共属一个主语insulation。so在此作为副词使用。

# Chapter 7
# Alternating-Current Electric Machines

## Part 1  Components and Principles of Operation of Induction Motors

Three-phase induction motors are the motors most frequently encountered in industry. They are simple, rugged, low-priced, and easy to maintain. They run at essentially constant speed from zero to full-load. The speed is frequency-dependent and consequently, these motors are not easily adapted to speed control. However, variable-frequency electronic drives are being used more and more to control the speed of commercial induction motors.

A 3-phase induction motor has two main parts: a stationary stator and a revolving rotor. The rotor is separated from the stator by a small air gap that ranges from 0.4mm to 4mm, depending on the power of the motor.*

The stator consists of a steel frame that supports a hollow, cylindrical core made up of stacked laminations. A number of evenly spaced slots, punched out of the internal circumference of the laminations, provide the space for the stator winding.

The rotor is also composed of punched laminations. These are carefully stacked to create a series of rotor slots to provide space for the rotor winding. We use two types of rotor windings: ① conventional 3-phase windings made of insulated wire and ② squirrel-cage windings. The type of winding gives rise to two main classes of motors: *squirrel-cage induction motors* (also called *cage motors*) and *wound-rotor induction motors*.

A squirrel-cage rotor is composed of bare copper bars, slightly longer than the rotor, which are pushed into the slots. The opposite ends are welded to two copper endrings so that all the bars are short-circuited together. As shown in Figure 7.1, the entire construction (bars and end-rings) resembles a squirrel cage, from which the name is derived.

The operation of a 3-phase induction motor is based upon the application of Faraday's Law and the Lorentz

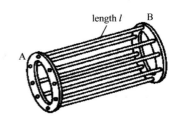

**Figure 7.1**  The entire construction of a squirrel-cage rotor

force on a conductor.

An induction machine carries alternating currents in both the stator and rotor windings. In a three-phase induction machine, the stator windings are connected to a balanced three-phase supply. The rotor windings are either short-circuited internally or connected through slip rings to a passive external circuit. The distinctive feature of the induction machine is that the rotor currents are induced by electromagnetic induction from the stator. This is the reason for the designation "induction machine".

The stator windings of a three-phase induction machine are similar to those of a synchronous machine. When balanced three-phase currents of frequency $f_s$ are applied, the stator windings produce a field rotating at synchronous speed given by

$$n_s = \frac{120 f_s}{p_f}$$

where $n_s$ is in r/min and $p_f$ is the number of poles (2 per three-phase winding set).

When there is a relative motion between the stator field and the rotor, voltages are induced in the rotor windings. The frequency ($f_r$) of the induced rotor voltages depends on the relative speeds of the start field and the rotor. The current in each rotor winding is equal to the induced voltage divided by the rotor circuit impedance at the rotor frequency $f_r$. The rotor current reacting with the stator field produces a torque that accelerates the rotor in the direction of the stator field rotation. As the rotor speed $n_r$ approaches the speed $n_s$ of the stator field, the induced rotor voltages and currents approach zero. To develop a positive torque, $n_r$ must be less than $n_s$. The rotor thus travels at a speed $n_s - n_r$ (r/min) in the backward direction with respect to the stator field. The *slip* speed of the rotor per unit of the synchronous speed is

$$s = \frac{n_s - n_r}{n_s}$$

The frequency $f_r$ of the induced rotor voltages is equal to the slip frequency $sf_s$.

At no load, the machine operates with negligible slip. If a mechanical load is applied, the slip increases (i.e., rotor speed decreases) such that the induced voltage and current produce the torque required by the load. Since there is a difference between the rotor speed and the synchronous speed, the induction machines are also called asynchronous machines. The machine thus operates as an *induction motor*.

Motors form a major portion of the system loads. Induction motors in particular form the workhorse of the electric power industry.

If the rotor is driven by a prime mover at a speed greater than that of the stator field, the slip is negative (i.e., rotor speed is greater than $n_s$). The polarities of the induced voltages are reversed so that the resulting torque is opposite in direction to that of rotation. The machine now operates as an *induction generator*.

# Chapter 7  Alternating-Current Electric Machines

## Part 2  Equivalent Circuit and Characteristics of Induction Motors

The equivalent circuit of the wound-rotor induction motor is shown in Figure 7.2. In this diagram, the circuit elements are fixed, except for the resistance $R_2/s$. Its value depends upon the slip and hence upon the speed of the motor.* Thus, the value of $R_2/s$ will vary from $R_2$ to infinity as the motor goes from start-up ($s = 1$) to synchronous speed ($s = 0$).

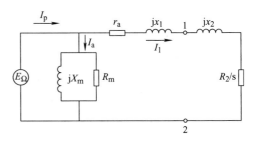

**Figure 7.2**  Equivalent Circuit of a wound-rotor motor referred to the primary (stator) side

This equivalent circuit of a wound-rotor induction motor is so similar to that of a transformer that it is not surprising that the wound-rotor induction motor is sometimes called a rotary transformer.*

The equivalent circuit of a squirrel-cage induction motor is the same, except that $R_2$ is then equal to the equivalent resistance $r_2$ of the rotor alone referred to as the stator, there being no external resistor.*

The torque developed by a motor depends upon its speed, but the relationship between the two cannot be expressed by a simple equation. Consequently, we prefer to show the relationship in the form of a curve. Figure 7.3 shows the torque-speed curve of a conventional 3-phase induction motor whose *nominal full-load torque* is T. The *starting torque* is 1.5T and the maximum torque (called *breakdown torque*) is 2.5T. *Pull-up torque* is the minimum torque developed by the motor while it is accelerating from rest to the breakdown torque.

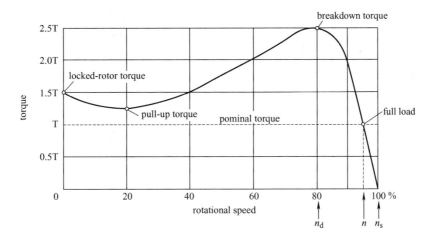

**Figure 7.3**  Typical torque-speed curve of a 3-phase squirrel-cage induction motor

At fullload, the motor runs at a speed of $n$. If the mechanical load increases slightly, the speed will drop until the motor torque is again equal to the load torque. As soon as the two torques are in balance, the motor will turn at a constant but slightly lower speed. However, if the load torque exceeds the breakdown torque, the motor will quickly stop. Small motors (15 hp and less) develop their breakdown torque at a speed of $n_d$ about 80% of synchronous speed. Big motors (1500 hp and more) attain their breakdown torque at about 98% of synchronous speed.

## Part 3  Synchronous Generators and Motors

Synchronous generators form the principal source of electric energy in power systems. Many large loads are driven by synchronous motors. Synchronous condensers are sometimes used as a means of providing reactive power compensation controlling voltage. These devices operate on the same principle and are collectively referred to as synchronous machines.

Commercial synchronous generators are built with either a stationary or a rotating DC magnetic field.

A *stationary-field synchronous generator* has the same outward appearance as a DC generator. The salient poles create the DC field, which is cut by a revolving armature. The armature possesses a 3-phase winding whose terminals are connected to three slip-rings mounted on the shaft. A set of bushes, sliding on the slip-rings, enables the armature to be connected to an external 3-phase load.

A *revolving-field synchronous generator* has a stationary armature called a stator. The 3-phase stator winding is directly connected to the load, without going through large, unreliable slip-rings and brushes. A stationary stator also makes it easier to insulate the windings because they are not subjected to centrifugal forces. Figure 7.4 is a schematic diagram of such a generator, sometimes called an alternator. The field is excited by a DC generator, usually mounted on the same shaft.

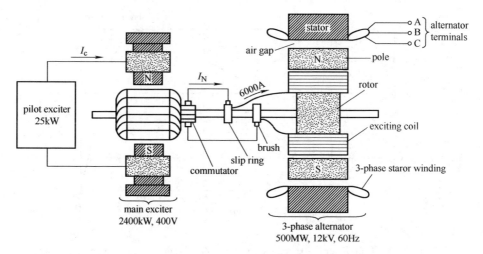

**Figure 7.4**  Schematic diagram and cross-section view of a typical 500 MW synchronous generator and its DC exciter

When the rotor is driven by a prime mover (turbine), the rotating magnetic field of the

# Chapter 7    Alternating-Current Electric Machines

field winding induces alternating voltages in the three-phase armature windings of the stator. The frequency of the induced alternating voltages and of the resulting currents that flow in the stator windings when a load is connected depends on the speed of the rotor.* The frequency of the stator electrical quantities is thus synchronized with the rotor mechanical speed; hence the designation "synchronous machine."

From an electrical standpoint, the stator of a synchronous generator is identical to that of a 3-phase induction motor. The winding is always connected in wye and the neutral is connected to ground. A wye connection is preferred to a delta connection because

(1) The voltage per phase is only 58% of the voltage between the lines. We can therefore reduce the amount of insulation in the slots which, in turn, enables us to increase the cross-section of the conductors which permits us to increase the current and, hence, the power output of the machine.*

(2) When a synchronous generator is under load, the voltage induced in each phase becomes distorted. With a wye connection, the distorting line-to-neutral undesired third harmonic does not appear between the lines because they effectively cancel each other.

Synchronous generators are built with two types of rotors: *salient-pole rotors* and smooth, *cylindrical rotors*. Salient-pole rotors are usually driven by low-speed hydraulic turbines, and cylindrical rotors are driven by high-speed steam turbines.

The quickness of response is one of the important features of the field excitation. In order to attain it, two DC generators are used: a main exciter and a pilot exciter. Static exciters that involve no rotating parts at all are also employed.

The main exciter feeds the exciting current to the field of the synchronous generator by way of brushes and slip-rings. It is regulated manually or automatically by control signals that vary the current $I_c$, produced by the pilot exciter, as shown in Figure 7.4.

The synchronous generator will deliver an increasing amount of reactive power to the system to which it is connected as its excitation is increased. Conversely, as its excitation is reduced it will furnish less reactive power and when underexcited will draw reactive power from the system.

When two or more synchronous machines are interconnected, the stator voltages and currents of all the machines must have the same frequency, and the rotor mechanical speed of each is synchronized to this frequency. Therefore, the rotors of all interconnected synchronous machines must be in synchronism.

As the name implies, synchronous motors run in synchronism with the revolving field. The speed of rotation is therefore tied to the frequency of the source. Because the frequency is fixed, the motor speed stays constant, irrespective of the load of the voltage of the 3-phase line. However, synchronous motors are used not so much because they run at constant speed but because they possess other unique electrical properties.*

Most synchronous motors are mainly used in heavy industry. There are some tiny single-phase synchronous motors used in control devices and electric clocks.

Synchronous Motors are identical in construction to salient-pole AC generators. The stator is composed of a slotted magnetic core, which carries a 3-phase lap winding. Consequently, the winding is also identical to that of a 3-phase induction motor. The rotor has a set of salient poles that are excited by a DC current. This damper winding serves to start the motor.

Modern synchronous motors often employ brushless excitation, as shown in Figure 7.5.

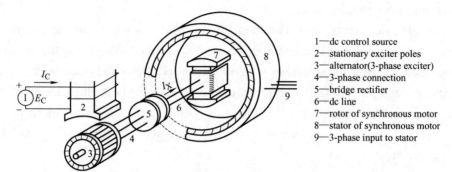

1—dc control source
2—stationary exciter poles
3—alternator(3-phase exciter)
4—3-phase connection
5—bridge rectifier
6—dc line
7—rotor of synchronous motor
8—stator of synchronous motor
9—3-phase input to stator

**Figure 7.5** Diagram showing the main components of a brushless exciter for a synchronous motor

A synchronous motor cannot start by itself; consequently, the rotor is usually equipped with a squirrel cage winding so that it can start up as an induction motor. When the stator is connected to the 3-phase line, the motor accelerates until it reaches a speed slightly below synchronous speed. The dc excitation is suppressed during this starting period.

Very large synchronous motors (20MW and more) are sometimes brought up to speed by an auxiliary motor, called a pony motor. Finally, in some big installations, the motor may be brought up to speed by a variable-frequency electronic source.

Owing to the inertia of the rotor and its load, a large synchronous motor may take several hours to stop after being disconnected from the line. To reduce the time, we use the following braking methods:

(1) Maintain full DC excitation with the armature in a short circuit.

(2) Maintain full DC excitation with the armature connected to three external resistors.

(3) Apply mechanical braking.

In methods (1) and (2), the motor slows down because it functions as a generator, dissipating its energy in the resistive elements of the circuit. Mechanical braking is usually applied only after the motor has reached half-speed or less. A lower speed prevents undue wear of the brake shoes.

## Part 4  The Synchronous Motor versus The Induction Motor

We have already seen that induction motors have excellent properties for speeds above 600r/min. But at lower speeds, they become heavy, costly, and have relatively low power factor, and efficiencies.

Synchronous motors are particularly attractive for low-speed drives because the power factor can always be adjusted to 1.0 and the efficiency is high. Although more complex to

# Chapter 7  Alternating-Current Electric Machines

build, their weight and cost are often less than those of induction motors of equal power and speed. This is particularly true for speeds below 300r/min.

A synchronous motor can improve the power factor of a plant while carrying its rated load. Furthermore, its starting torque can be considerably greater than that of an induction motor. The reason is that the resistance of the squirrel-cage winding can be high without affecting the speed or efficiency at synchronous speed. Figure 7.6 compares the properties of a squirrel-cage induction motor and a synchronous motor having the same nominal rating. The biggest difference is in the starting torque. High-power electronic converters generating very low frequencies enable us to run synchronous motors at ultra-low speeds. Thus, huge motors in the 10MW range drive crushers, rotary kilns, and variable-speed ball mills.

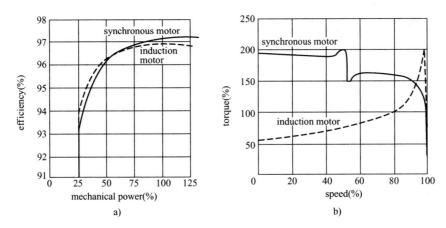

**Figure 7.6**  Comparison between the efficiency and starting torque of a squirrel-cage induction motor and a synchronous motor, both rated at 4000 hp, 1800r/min, 6.9kV, 60Hz

## NEW WORDS AND EXPRESSIONS

| | | | |
|---|---|---|---|
| 1. | rugged | adj. | 朴实的；强壮的 |
| 2. | fullload | | 满负荷 |
| 3. | stationary | adj. | 固定的 |
| 4. | stator | n. | 定子 |
| 5. | revolving | adj. | 旋转的 |
| 6. | rotor | n. | 转子 |
| 7. | air gap | | 气隙 |
| 8. | cylindrical | adj. | 圆柱的 |
| 9. | lamination | n. | 迭片结构 |
| 10. | evenly | adv | 均匀地 |
| 11. | slot | n. | 缝，狭槽 |
| 12. | punch | vt. | 冲压 |
| 13. | squirrel-cage | | 鼠笼，关松鼠等的笼子 |
| 14. | wound-rotor | | 绕线转子 |
| 15. | weld | vt. | 焊接 |

| | | | |
|---|---|---|---|
| 16. | short-circuit | | 使短路，缩短 |
| 17. | Lorentz force | | 洛伦兹力 |
| 18. | slip ring | *n.* | 滑动环；集电环 |
| 19. | passive | *adj.* | 无源的 |
| 20. | electromagnetic | *adj.* | 电磁的 |
| 21. | pole | *n.* | 磁极，电极 |
| 22. | relative motion | | 相对运动 |
| 23. | stator field | | 定子磁场 |
| 24. | induce | *vt.* | 感应 |
| 25. | torque | *n.* | 扭矩，转矩 |
| 26. | slip | *n.* | 转差率 |
| 27. | asynchronous | *adj.* | 异步的 |
| 28. | workhorse | *n.* | 重负荷机器 |
| 29. | polarity | *n.* | 极性 |
| 30. | synchronous condenser | | 同步调相机 |
| 31. | salient | *adj.* | 突出的，易见的，显著的 |
| 32. | armature | *n.* | 电枢（电机的部件） |
| 33. | shaft | *n.* | 轴，杆状物 |
| 34. | centrifugal force | | 离心力 |
| 35. | alternator | *n.* | 交流发电机 |
| 36. | exciter | *n.* | 励磁机 |
| 37. | wye | *n.* | Y字，Y字形物 |
| 38. | distorted | *adj.* | 畸变的，扭歪的 |
| 39. | cylindrical | *adj.* | 圆柱的，圆筒形的 |
| 40. | pilot exciter | | 副励磁机 |
| 41. | lap winding | | 叠绕组 |
| 42. | damper winding | | 阻尼绕组 |
| 43. | brushless | *adj.* | 不带电刷的 |
| 44. | auxiliary | *adj.* | 辅助的 |
| 45. | pony motor | | 辅助电动机，起动[伺服]电动机 |
| 46. | inertia | *n.* | 惯性，惯量 |
| 47. | brake | *v.* | 刹车，制动 |
| 48. | brake shoe | | 闸皮 |
| 49. | versus | *prep.* | 与…相对 |
| 50. | crusher | *n.* | 轧碎机 |
| 51. | rotary kiln | | 转炉 |
| 52. | ball mill | *n.* | 球磨机 |

## PHRASES

### 1. adapted to: 适合

Example: The speed is frequency-dependent and consequently, these motors are not easily <u>adapted to</u> speed control.

# Chapter 7  Alternating-Current Electric Machines

速度是由频率决定的，因此，这些转子不适合进行速度控制。

**2. give rise to: 引起，使发生**

Example: The type of winding gives rise to two main classes of motors: squirrel-cage induction motors (also called cage motors) and wound-rotor induction motors.

该绕组类型引出两个主要类别的转子：鼠笼式感应电动机（也叫笼型感应电动机）和绕线式感应电动机。

**3. in the form of: 以…的形式**

Example: Consequently, we prefer to show the relationship in the form of a curve.
因此，我们更喜欢以曲线的形式显示其关系。

**4. as the name implies: 正如名字所暗示的，顾名思义**

Example: As the name implies, synchronous motors run in synchronism with the revolving field.
顾名思义，同步电动机与旋转磁场同步运行。

**5. tie to: 依靠，依赖**

Example: The speed of rotation is therefore tied to the frequency of the source.
因此，转速取决于电源的频率。

## COMPLICATED SENTENCES

**1. The rotor is separated from the stator by a small air gap that ranges from 0.4mm to 4mm, depending on the power of the motor.**

译文：转子与定子通过气隙隔开，该气隙从 0.4mm 到 4mm，这取决于电动机的功率。

说明：a small air gap 作为介词的宾语，本身又带有较长的定语，这在翻译时容易造成句子拗口不通顺，可以适当拆解和调整语序，构造新的句子，把较长的定语以谓语的形式表示。

The rotor is separated from the stator by a small air gap; the air gap ranges from 0.4mm to 4mm; depending on the power of the motor.

**2. Its value depends upon the slip and hence upon the speed of the motor.**

译文：其数值取决于转差率，因而由电动机的速度决定。

说明：句中出现了省略的情况。事实上，从逻辑上讲，连词 and 连接了两个并列的谓语，分别是 depends upon the slip 和 depends upon the speed of the motor，其中第二个谓语的动词 depend 省略了，在翻译时应注意。

**3. This equivalent circuit of a wound-rotor induction motor is so similar to that of a transformer that it is not surprising that the wound-rotor induction motor is sometimes called a rotary transformer.**

译文：绕线式感应电动机的这个等效电路与变压器的如此类似，以至绕线式感应电动机有时被称为旋转变压器并不奇怪。

说明：词组 be similar to 表示"与…相似"。句中的三个 that 用法各不相同。第一个 that 引导了一个名词短语 that of a transformer，代指 equivalent circuit，作为 be similar to 的宾语。第二个 that 与 so 遥相呼应，组成一个词组 so ... that...，表示"如此…以至…"。第三个与前面的内容构成句型 it is not surprising that，实际上它是引导了一个宾语从句。为了便于理解，这个句子可以拆解为：

This equivalent circuit of a wound-rotor induction motor is similar to that of a transformer, so it is not surprising that the wound-rotor induction motor is sometimes called a rotary transformer.

**4. The equivalent circuit of a squirrel-cage induction motor is the same, except that $R_2$ is then equal to the equivalent resistance $r_2$ of the rotor alone <u>referred to as the stator</u>, there being no external resistor.**

译文：笼型感应电动机的等效电路是一样的，不过 $R_2$ 等于折算到定子侧的转子自己的等效电阻 $r_2$，没有额外的电阻。

说明：词组 refer to 表示"折算到"。介词 except 有两个宾语，一是 that 从句，一是 there be 的动名词短语。

**5. The frequency of the induced alternating voltages and of the resulting currents that flow in the stator windings when a load is connected depends on the speed of the rotor.**

译文：感应出的交流电压的频率和接有负荷时流过定子的电流的频率，取决于转子的速度。

说明：句中出现了省略的情况。事实上，连词 and 连接了两个并列的主体，共同作为整个句子的主语，共用一个谓语动词 depend。这两个主体分别是 the frequency of the induced alternating voltages 和 the frequency of the resulting currents，其中第二个短语的 frequency 省略了，在翻译时应注意。

**6. We can therefore reduce the amount of insulation in the slots which, in turn, enables us to increase the cross-section of the conductors which permits us to increase the current and, hence, the power output of the machine.**

译文：因此，我们可以减少槽中的绝缘（介质）数量，这样可以增大导体的横截面，从而提高电流，进而提高电机的输出功率。

说明：句子中出现了含义接近、前后重复的内容 enables us to 和 permits us to，在翻译时可以适当省略，以使句子简单明了。

**7. However, synchronous motors are used not so much because they run at constant speed but because they possess other unique electrical properties.**

译文：然而，使用同步电动机不仅仅是因为它们以恒定的速度运行，还由于它们具有其他独特的电气特性。

说明：这个句子翻译时很容易出错。关键在于 are used not so much because 部分如何理解。如果拆解成 are used not so much, because，则句子的意思是"同步电动机用

# Chapter 7　Alternating-Current Electric Machines

的不多，这是因为"，与后面的 but because 不匹配。如果拆解成 are <u>used so much</u> not because，则句子的意思是"同步电动机用的很多，不是因为"，与后面的 but because 可以匹配，但是从上下文看不到任何地方介绍"同步电动机用的<u>很多</u>"。所以比较合理的拆解方法是 are used, <u>not so much</u> because，意思是"同步电动机被使用，不是那样多的因为"。

## ABBREVIATIONS (ABBR.)

hp　　　horsepower　　　　　　　　　　　　　马力

## SUMMARY OF GLOSSARY

1. 交流电机　alternating machines
    induction machine　　　　　　　　感应电机
    asynchronous machine　　　　　　异步电机
    synchronous machine　　　　　　　同步电机
2. 感应电动机　induction motors
    wound-rotor induction motors　　　绕线式感应电动机
    squirrel-cage induction motors　　　鼠笼式（笼型）感应电动机
    cage motors　　　　　　　　　　　鼠笼式电机
3. 电动机转矩　torques of motors
    nominal full-load torque　　　　　　标称满载转矩
    starting torque　　　　　　　　　　起动转矩
    breakdown torque　　　　　　　　　停转力矩，极限转矩，崩溃转矩
    pull-up torque　　　　　　　　　　拉起转矩，最低起动转矩
5. 速度　speed
    synchronous speed　　　　　　　　同步速
    slip speed　　　　　　　　　　　　转差速度
    rotor speed　　　　　　　　　　　转子速度
6. 同步发电机　synchronous generators
    stationary-field synchronous generator　　静止磁场同步发电机
    revolving-field synchronous generator　　旋转磁场同步发电机
    alternator　　　　　　　　　　　　交流发电机（旋转磁场同步机）
7. 同步机转子　rotors of synchronous generators
    salient-pole rotors　　　　　　　　凸极转子
    cylindrical rotors　　　　　　　　　鼓形转子
8. 励磁机　exciters
    main exciter　　　　　　　　　　　主励磁机
    pilot exciter　　　　　　　　　　　副励磁机
    static exciter　　　　　　　　　　　静态励磁机

# EXERCISES

**1. Translate the following words or expressions into Chinese.**

(1) alternator　　　　　(2) stator　　　　　　(3) squirrel-cage
(4) slip ring　　　　　 (5) armature　　　　　(6) relative motion
(7) stator field　　　　(8) asynchronous　　　(9) Lorentz force
(10) pilot exciter　　　(11) brake　　　　　　(12) salient-pole

**2. Translate the following words or expressions into English.**

(1) 满负荷　　　　　　　(2) 转子　　　　　　　(3) 绕线式转子
(4) 转矩　　　　　　　　(5) 主励磁机　　　　　(6) 气隙
(7) 起动 [ 伺服 ] 电动机　(8) 无电刷的　　　　　(9) 滑差（转差率）

**3. Fill in the blanks with proper words or expressions.**

(1) The type of winding gives rise to two main classes of motors: _____ induction motors (also called cage motors) and _____ induction motors.

(2) This equivalent circuit of a wound-rotor induction motor is so similar to that of a transformer that it is not surprising that the wound-rotor induction motor is sometimes called a _____ transformer.

(3) Since there is a difference between the rotor speed and the synchronous speed, the induction machines are also called _____ machines.

(4) Commercial synchronous generators are built with either a _____ or a _____ DC magnetic field.

# Word-Building (7)　-wise, -ward　方向，方式

**1. -wise 副词后缀，表示方向，方式**

| clock | clockwise | ad. | 顺时针方向地（的） |
| length | lengthwise | ad. | 纵向地（的）；沿长度方向（的） |
| like | likewise | adv. | 同样地，照样地 |
| other | otherwise | ad. | 另外（的）；其他方面（的） |
| step | stepwise | ad. | 逐步（的），逐渐（的），分步（的） |

**2. -ward 副词后缀，表示方向，方式**

| back | backward | ad. | 向后地（的），相反地（的） |
| for | forward | ad. | 前进（的），向前（的）； |
| in | inward | ad. | 向内（的），内在（的） |
| out | outward | ad. | 外面的，外表的；公开，向外 |

# Chapter 8
# Direct-Current Electric Machines

## Part 1  Generating an AC Voltage

Irrelevant as it may seem, the study of a DC generator has to begin with a knowledge of the AC generator.* The reason is that the voltage generated in any DC generator is inherently alternating and only becomes DC after it has been rectified by the commutator.

Figure 8.1 shows an elementary AC generator composed of a coil that revolves at 60r/min between the N and S poles of a permanent magnet. The rotation is due to an external driving force, such as a motor (not shown). The coil is connected to two slip rings mounted on the shaft. The slip rings are connected to an external load by means of two stationary brushes x and y.

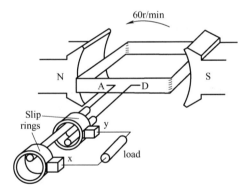

**Figure 8.1**  Schematic diagram of an elementary AC generator turning at 1 revolution per second

As the coil rotates, a voltage is induced between its terminals A and D. This voltage appears between the brushes and, therefore, across the load. The voltage is generated because the conductors of the coil <u>cut across</u> the flux produced by the N and S poles. The induced voltage is, therefore, maximum (20V, say) when the coil is momentarily in the horizontal position, as shown. No flux is cut when the coil is momentarily in the vertical position; consequently, the voltage at these instants is zero. Another feature of the voltage is that its polarity changes every time the coil makes half a turn. The voltage can therefore be represented as a function of the angle of rotation (Figure 8.2). The wave shape depends upon the shape of the N and S poles. We assume the poles were designed to generate the sinusoidal wave shown.

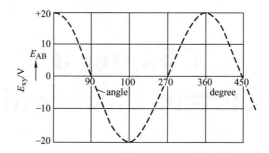

**Figure 8.2** Voltage induced in the AC generator as a function of the angle of rotation

The coil in our example revolves at a uniform speed; therefore each angle of rotation corresponds to a specific interval of time. Because the coil makes one turn per second, the angle of 360° in Figure 8.2 corresponds to an interval of one second.

## Part 2　Direct-Current Generator

If the brushes in Figure 8.1 could be switched from one slip ring to the other every time the polarity was about to change, we would obtain a voltage of constant polarity across the load.* Brush x would always be positive and brush y negative. We can obtain this result by using a commutator (Figure 8.3). A commutator in its simplest form is composed of a slip ring that is cut in half, with each segment insulated from the other as well as from the shaft*. One segment is connected to coil-end A and the other to coil-end D. The commutator revolves with the coil and the voltage between the segments is picked up by two stationary brushes x and y.

The voltage between brushes x and y pulsates but never changes polarity (Figure 8.4). The alternating voltage in the coil is rectified by the commutator, which acts as a mechanical reversing switch.

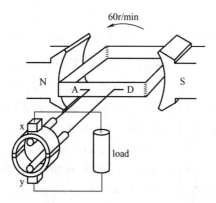

**Figure 8.3** An elementary DC generator is simply an AC generator equipped with a mechanical rectifier called a commutator

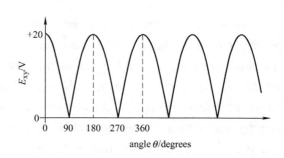

**Figure 8.4** The elementary DC generator produces a pulsating DC voltage

# Chapter 8  Direct-Current Electric Machines

The elementary AC and DC generators in Figure 8.1 and Figure 8.3 are essentially built the same way. In each case, a coil rotates between the poles of a magnet, and an AC voltage is induced in the coil. The machines only differ in the way the coils are connected to the external circuit (Figure 8.5: AC generators carry slip rings (Figure 8.5b) while DC generators require a commutator (Figure 8.5a). We sometimes build small machines which carry both slip rings and a commutator (Figure 8.5c). Such machines can function simultaneously as AC and DC generators.

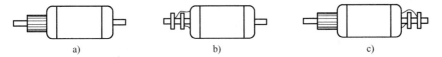

**Figure 8.5**  The three armatures have identical windings. Depending upon how they are connected (to slip rings or a commutator), an AC or DC voltage is obtained

The voltage still pulsates but it never falls to zero; it is much closer to a steady DC voltage. By increasing the number of coils and segments, we can obtain a DC voltage that is very smooth. Modern DC generators produce voltages having a ripple of less than 5 percent. The coils are lodged in the slots of a laminated iron cylinder. The coils and the cylinder constitute the armature of the machine. The percent ripple is the ratio of the RMS value of the AC component of voltage to the DC component, expressed in percent.

The current in the external load always flows in the same direction. The machine represented in Figure 8.3 is called a direct-current generator, or dynamo.

Until now, we have assumed that the only magnetomotive force (MMF) acting in a DC generator is due to the field. However, the current flowing in the armature coils also creates a powerful magnetomotive force that distorts and weakens the flux coming from the poles. This distortion and field weakening takes place in both motors and generators. The effect produced by the armature MMF is called armature reaction.

The equivalent circuit of a DC generator is thus composed of a resistance $R_o$ in series with a voltage $E_o$ (Figure 8.6). The latter is the voltage induced in the revolving conductors. Terminals 1 and 2 are the external armature terminals of the machine, and $F_1$-$F_2$ are the field winding terminals.

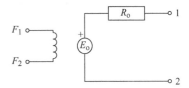

**Figure 8.6**  Equivalent circuit of a dc generator

## Part 3  Direct-Current Motors

Commercial DC generators and motors are built the same way, consequently, any DC generator can operate as a motor and vice versa. Owing to their similar construction, the fundamental properties of generators and motors are identical. Consequently, anything we learn about a DC generator can be directly applied to a DC motor.

To illustrate, consider a DC generator in which the armature, initially at rest, is connected to a DC source $E_s$ by means of a switch.*

As soon as the switch is closed, a large current flows in the armature because its resistance is very low. The individual armature conductors are immediately subjected to a force because they are immersed in the magnetic field created by the permanent magnets. These forces add up to produce a powerful torque, causing the armature to rotate.

On the other hand, as soon as the armature begins to turn, a second phenomenon takes place: the generator effect. We know that a voltage $E_o$ is induced in the armature conductors as soon as they cut a magnetic field. This is always true, no matter what causes the rotation. The value and polarity of the induced voltage are the same as those obtained when the machine operates as a generator. The induced voltage $E_o$ is therefore proportional to the speed of rotation n of the motor and the flux $\Phi$ per pole, as previously given:

$$E_o = Z_n \Phi / 60 \qquad (8.1)$$

It acts against the voltage in the sense that the net voltage acting in the series circuit of Figure 8.7 is equal to $(E_s - E_o)$ volts and not $(E_s + E_o)$ volts. The resulting armature current $I$ is limited only by the armature resistance R, and so

$$I = (E_s - E_o)/R \qquad (8.2)$$

When the motor is at rest, the induced voltage $E_o = 0$, and so the starting current is

$$I = E_s/R \qquad (8.3)$$

The starting current may be 20 to 30 times greater than the nominal full-load current of the motor. In practice, this would cause the fuses to blow or the circuitbreakers to trip. However, if they are absent, the large forces acting on the

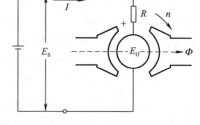

**Figure 8.7** Counter-electromotive force in a dc motor

armature conductors produce a powerful starting torque and a consequent rapid acceleration of the armature.

As the speed increases, the counter-emf (electromotive force) $E_o$ increases, with the result that the value of $(E_s - E_o)$ diminishes. It follows from Eq. 8.2 that the armature current $I$ drops progressively as the speed increases.

Although the armature current decreases, the motor continues to accelerate until it reaches a definite, maximum speed. At no-load this speed produces a counter-emf $E_o$ slightly less than the source voltage $E_s$. In effect, if $E_o$ were equal to $E_s$ the net voltage $(E_s - E_o)$ would become zero and so, too would the current $I$.* The driving forces would cease to act on the armature conductors, and the mechanical drag imposed by the fan and the bearings would immediately cause the motor to slow down. As the speed decreases the net voltage $(E_s - E_o)$ increases and so does the current $I$. The speed will cease to fall as soon as the torque developed by the armature current is equal to the load torque. Thus, when a motor runs at no-load, the counter-emf must be slightly less than $E_s$, so as to enable a small current to flow, sufficient to produce the required torque.

# Chapter 8  Direct-Current Electric Machines

The mechanical power and torque of a DC motor are two of its most important properties.

When a large DC motor is coupled to a heavy inertia load, it may take an hour or more for the system to come to a halt. For many reasons such a lengthy deceleration time is often unacceptable and, under these circumstances, we must apply a braking torque to ensure a rapid stop. One way to brake the motor is by simple mechanical friction, in the same way we stop a car. A more elegant method consists of circulating a reverse current in the armature, so as to brake the motor electrically. Two methods are employed to create such an electromechanical brake: ① dynamic braking and ② plugging.

## Part 4  The Application of DC Machines

Direct-current generators are not as common as they used to be, because direct current, when required, is mainly produced by electronic rectifiers.* Indeed, many DC motors in the industry operate as generators for brief periods.

Direct-current motors drive devices such as hoists, fans, pumps, calenders, punch-presses, and cars. These devices may have a definite torque-speed characteristic (such as a pump or fan) or a highly variable one (such as a hoist or automobile). The torque-speed characteristic of the motor must be adapted to the type of load it has to drive.

Direct-current motors are seldom used in ordinary industrial applications because all electric utility systems furnish alternating current. However, for special applications such as steel mills, mines, and electric trains, it is sometimes advantageous to transform the alternating current into direct current in order to use DC motors. The reason is that the torque-speed characteristics of dc motors can be varied over a wide range while retaining high efficiency.

Today, this general statement can be challenged because the availability of sophisticated electronic drives has made it possible to use alternating current motors for variable speed applications. Nevertheless, there are millions of DC motors still in service and thousands more are being produced every year.

## NEW WORDS AND EXPRESSIONS

| | | | |
|---|---|---|---|
| 1. | rectify | vt. | 整流 |
| 2. | commutator | n. | 换向器，转接器 |
| 3. | permanent magnet | | 永磁体 |
| 4. | driving force | | 驱[传，主]动力 |
| 5. | uniform speed | | 匀速，等速 |
| 6. | pulsate | vi. | 搏动，跳动 |
| 7. | reversing switch | | 换向开关 |

| 8. | ripple | n. | 波纹；纹波 |
| 9. | lodge | vt. | 容纳；寄存；把…射入，存放 |
| 10. | RMS | n. | 均方根；有效值 |
| 11. | dynamo | n. | 发电机 |
| 12. | armature reaction | | 电枢反应 |
| 13. | cease | v. | 停止，终了 |
| 14. | no-load | n. | 空载 |
| 15. | elegant | adj. | 讲究的；优雅的 |
| 16. | plugging | n. | 反向制动 |
| 17. | hoist | n. | 起重机 |
| 18. | calender | n. | 砑光机 |
| 19. | punch-press | | 冲压机 |
| 20. | electric utility | | 电业 |
| 21. | mill | n. | 工厂，制造厂；压榨机，磨坊 |
| 22. | mine | n. | 矿，矿山 |

## PHRASES

**1. cut across: 愿意是"抄近路通过"，在这里指"切割"**

Example: The voltage is generated because the conductors of the coil <u>cut across</u> the flux produced by the N and S poles.

线圈的导体切割 N 极和 S 极产生的磁通，从而产生了电压。

**2. corresponds to: 等于，相当于，与…相似**

Example: Because the coil makes one turn per second, the angle of 360° in Figure 8.2 <u>corresponds to</u> an interval of one second.

由于线圈每秒钟转一圈，图 8-2 中 360°的角相当于 1s 的时间间隔。

**3. vice versa: adv. 反之亦然**

Example: Commercial dc generators and motors are built the same way; consequently, any dc generator can operate as a motor and <u>vice versa</u>.

商业化的直流发电机和电动机同样制造，因此，直流发电机可以作为电动机来运行，反之亦然。

**4. be subjected to: 服从，遭受**

**be immersed in: 陷入，陷于**

Example: The individual armature conductors <u>are</u> immediately <u>subjected to</u> a force because they <u>are immersed in</u> the magnetic field created by the permanent magnets.

单个的电枢导体会立即受到一个力，因为它们浸陷于永磁体所产生的磁场之中。

# Chapter 8　Direct-Current Electric Machines

**5. in the sense that:** 从……意义上

**act against:** 违反

Example: It acts against the voltage <u>in the sense</u> that the net voltage acting in the series circuit of Figure 8.7 is equal to ($E_s - E_o$) volts and not ($E_s + E_o$) volts.

从某种意义上说，它与该电压相反，即在图 8-7 所示的串联电路中的净电压等于 ($E_s - E_o$) 而不等于 ($E_s + E_o$)。

**6. in the same way:** 以和…相同的方法

Example: One way to brake the motor is by simple mechanical friction, <u>in the same way</u> we stop a car.

一种用于电动机制动的方法是通过简单的机械摩擦，就像我们停车那样。

## COMPLICATED SENTENCES

**1. Irrelevant as it may seem, the study of a DC generator has to begin with <u>a knowledge of</u> the AC generator.**

译文：可能看上去并不相关，直流发电机的研究必须从对交流发电机的了解开始。

说明：句首的分句采用了倒装的形式，表示对提到前面的内容的突出和强调。词组 a knowledge of 的意思是"对……的了解"，这里作"认识，了解，熟悉"解，是可数名词。词组 begin with 有两种用法，一是作状语，相当于副词短语，意思是"首先"；一是作为动词词组，表示"从……开始，以……为开端"。这里 begin with 的意思为后者。

**2. If the brushes in Figure 8.1 could be switched from one slip ring to the other every time the polarity <u>was about to</u> change, we would obtain a voltage of constant polarity across the load.**

译文：如果每次极性将要改变时图 8-1 中的电刷都能从一个集电环切换到另一个（集电环），就可以得到经过负荷的极性不变的电压。

说明：词组 every time 引导了一个时间状语从句，在翻译时可以适当提前（按照中文的表达习惯，时间状语一般不出现在谓语动词的后面）。词组 be about to 表示"即将，将要，打算"。英文科技文献习惯用被动语态的形式表达，当不方便用被动语态时，往往用 we 作为主语构造主动语态的句子，而作为主语的 we 通常不必翻译。

**3. A commutator in its simplest form is composed of a slip ring that is cut in half, with each segment insulated from the other as well as from the shaft.**

译文：形式最简单的换向器由一分为二的集电环组成，每一段在与轴绝缘的同时也彼此绝缘。

说明：词组 in its simplest form 意思是"以其最简单的形式"。词组 cut in half 意思是"切为两半"。词组 as well as 表示"在…的同时也…"，需要注意的是，如果以（State A）as well as（State B）的形式出现，则恰当的翻译方法是"在（State B）的同时也（State A）"。

**4. To illustrate, consider a DC generator in which the armature, initially at rest, is connected to a DC source $E_s$ by means of a switch.**

译文：为了说明，考虑一个定子最初静止、通过开关与直流电源 $E_s$ 相连的直流发电机。

说明：由 in which 引导的定语从句修饰句子的宾语 a dc generator。在定语从句中，initially at rest 是 is initially at rest 的省略写法。

**5. In effect, if $E_o$ were equal to $E_s$ the net voltage ($E_s - E_o$) would become zero and so, too would the current $I$.**

译文：从效果上，如果 $E_o$ 等于 $E_s$，则净电压 ($E_s - E_o$) 将变为零，并且电流 $I$ 也同样。

说明：in effect 若按词组翻译，意思为"有效（地）"，在句子中讲不通。事实上，若其按各自的本义翻译，意思为"在效果上"，还是可以接受的。后面的内容用了以 so 引导的倒装形式，省略了一些内容。如果调整为正常的语序，并补全省略掉的词语，则应为：

... and the current $I$ would become zero too.

**6. Direct-current generators are not as common as they used to be, because direct current, when required, is mainly produced by electronic rectifiers.**

译文：直流发电机用的不如过去那么多了，这是因为需要的时候直流主要由电子整流器产生。

说明：句子中的 when required 作为插入语，在翻译的时候可以提到主语的前面。

## ABBREVIATIONS (ABBR.)

| | | | |
|---|---|---|---|
| 1. | RMS | root mean square | 均方根 |
| 2. | MMF | magneto motive force | 磁动势，磁通势 |
| 3. | EMF | electro motive force | 电动势 |

## SUMMARY OF GLOSSARY

1. **直流发电机　direct-current generators**
   
   | | |
   |---|---|
   | direct-current generator | 直流发电机 |
   | dc generator | 直流发电机 |
   | dynamo | （直流）发电机 |

2. **感应电动机　induction motors**
   
   | | |
   |---|---|
   | mechanical braking | 机械制动 |
   | electromechanical braking | 机电制动 |
   | dynamic braking | 动态制动 |
   | plugging | 反向制动（堵塞制动） |

# Chapter 8  Direct-Current Electric Machines

## EXERCISES

**1. Translate the following words or expressions into Chinese.**

(1) driving force        (2) rectify        (3) mechanical braking
(4) RMS value          (5) dynamo        (6) electric utility
(7) commutator         (8) fuse to blow    (9) net voltage

**2. Translate the following words or expressions into English.**

(1) 换向开关            (2) 纹波            (3) 电枢反应
(4) 空载               (5) 永磁体           (6) 减速
(7) 反向制动            (8) 磁动势           (9) 整流器

**3. Fill in the blanks with proper words or expressions.**

(1) Commercial dc generators and _____ are built the same way; consequently, any dc generator can operate as a _____ and vice versa.

(2) Direct-current generators are not as common as they used to be, because direct current, when required, is mainly produced by _____.

(3) Two methods are employed to create such an electromechanical brake: _____ and _____.

# Word-Building (8)　in-, ir-　否定 2

**1. in-　前缀，表示"否定，与…相反"之义，可译为：不，非**

| | | | |
|---|---|---|---|
| accurate | inaccurate | *adj.* | 不准确的，不精确的 |
| correct | incorrect | *adj.* | 不正确的，错误的 |
| correlate | incorrelate | *adj.* | 不相关的 |
| dependent | independent | *adj.* | 不受约束的，独立自主的 |
| equality | inequality | *n.* | 不平等，不同，不等式 |
| effective | ineffective | *adj.* | 无效的，工作效率低的 |
| evitable | inevitable | *adj.* | 不可避免的 |
| exhaustible | inexhaustible | *adj.* | 无穷尽的，用不完的 |
| expensive | inexpensive | *adj.* | 不贵重的，便宜的 |
| flexible | inflexible | *adj.* | 不灵活的，不可变更的 |
| variable | invariable | *adj.* | 不变的，不变量 |
| expensive | inexpensive | *adj.* | 不贵重的，便宜的 |

**2. ir-　前缀，与 in-，im- 相同，表示"否定，与…相反"之义（用于 r 之前）**

| | | | |
|---|---|---|---|
| respective | irrespective | *adj.* | 不顾的，不考虑的，无关的 |
| rational | irrational | *adj.* | 无理性的，失去理性的 |
| realizable | irrealizable | *adj.* | 不能实现的，不能达到的 |

| | | | |
|---|---|---|---|
| reciprocal | irreciprocal | *adj.* | 非交互的，非互惠的 |
| recognizable | irrecognizable | *adj.* | 不能识别的 |
| reconcilable | irreconcilable | *adj.* | 不能协调的，矛盾的 |
| recoverable | irrecoverable | *adj.* | 无可挽救的 |
| regular | irregular | *adj.* | 不规则的，无规律的 |
| relative | irrelative | *adj.* | 无关系的 |
| relevant | irrelevant | *adj.* | 不相关的，不切题的 |

# Chapter 9
# Switch Devices

## Part 1  Circuit Breakers

A circuit breaker (CB) can be described as a device used in an electrical network to ensure the uninterrupted flow of current in that network under normal operating conditions, and to interrupt the flow of excessive current in a faulty network. Under some circumstances, it may also be required to interrupt load current and to perform open-close-open sequences (auto reclosing) on a fault on others. The successful achievement of these, and indeed all duties rely on extremely careful design, development, and proving tests on any circuit breaker. A good and reliable mechanical design is required to meet the demands of opening and closing the circuit breaker contacts, and an effective electrical design is also essential to ensure that the CB can deal with any of the electrical stresses encountered in service.

Circuit breakers are designed to interrupt either the mid or short-circuit currents. They behave like big switches that may be opened or closed by local pushbuttons or by distant telecommunication signals emitted by the system of protection. Thus, circuit breakers will automatically open a circuit whenever the line current, line voltage, frequency, and so on, <u>depart from</u> a preset limit.

The nameplate on a circuit breaker usually indicates ① the maximum steady-state current it can carry, ② the maximum interrupting current, ③ the maximum line voltage, and ④ the interrupting time in cycles. The interrupting time may last from 3 to 8 cycles on a 60Hz system. To interrupt large currents this quickly, we have to ensure rapid deionization of the arc, combined with rapid cooling. High-speed interruption limits the damage to transmission lines and equipment and, equally important, it helps to maintain the stability of the system whenever a contingency occurs.

The triggering action that causes a circuit breaker to open is usually produced by means of an overload relay that can detect abnormal line conditions. For example, the relay coil in Figure 9.1 is connected to the secondary of a current transformer. The primary carries the line current exceeds a preset limit, the secondary current will cause relay contacts $C_1$ and $C_2$ to close. As soon as they close, the tripping coil is energized by an auxiliary DC source. This causes the three main line contacts to open, thus interrupting the circuit.

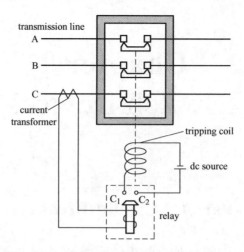

**Figure 9.1** Elementary tripping circuit for a circuit breaker

The most important types of circuit breakers are the following:

### 1. Oil Circuit Breakers (OCBs)

Oil circuit breakers are composed of a steel tank filled with insulating oil. In one version (Figure 9.2), three porcelain bushings channel the 3-phase line currents to a set of fixed contacts. Three movable contacts, actuated simultaneously by an insulated rod, open and close the circuit.* When the circuit breaker is closed, the line current for each phase penetrates the tank by way of a porcelain bushing, flows through the first fixed contact, the movable contact, the second fixed contact, and then on out by a second bushing.*

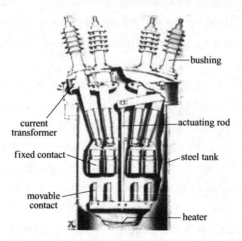

**Figure 9.2** Cross-section of an oil circuit breaker. Showing four of the six bushings; the heater keeps the oil at a satisfactory temperature during cold weather

If an overload occurs, the tripping coil releases a powerful spring that pulls on the insulated rod, causing the contacts to open. As soon as the contacts separate, a violent arc is created, which volatilizes the surrounding oil. The pressure of the hot gases creates turbulence around the contacts. This causes cool oil to swirl around the arc, thus extinguishing it.

# Chapter 9　Switch Devices

## 2. Air-Blast Circuit Breakers

These circuit breakers interrupt the circuit by blowing compressed air at supersonic speed across the opening contacts. Compressed air is stored in reservoirs at a pressure of about 3MPa and is replenished by a compressor located in the substation. The most powerful circuit breakers can typically open short-circuit currents of 40kA at a line voltage of 765kV in a matter of 3 to 6 cycles on a 60Hz line. The noise accompanying the air blast is so loud that noise-suppression methods must be used when the circuit breakers are installed near residential areas. Figure 9.3 shows a cross-section of the contact module.

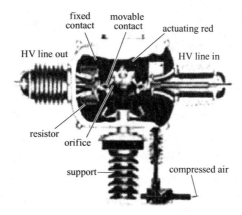

**Figure 9.3**　Cross-section of one module of an air-blast circuit breaker

## 3. $SF_6$ Circuit Breakers

These totally enclosed circuitbreakers, insulated with $SF_6$ gas, are used whenever space is at a premium, such as in down-town substations.* These circuit breakers are much smaller than any other type of circuit breaker of equivalent power and are far less noisy than air circuit breakers.

## 4. Vacuum Circuit Breakers

These circuit breakers operate on a different principle from other breakers because there is no gas to ionize when the contacts open. They are hermetically sealed; consequently, they are silent and never become polluted. Their interrupting capacity is limited to about 30kV. For higher voltages, several circuit breakers are connected in series. Vacuum circuit breaks are often used in underground distribution systems.

# Part 2　Disconnecting Switches

Air-break switches can interrupt the exciting currents of transformers or the moderate capacitive currents of unloaded transmission lines. They cannot interrupt normal load currents.

Air-break switches are composed of a movable blade that engages a fixed contact; both are mounted on insulating supports (Figure 9.4).* Two arcing horns are attached to the fixed and movable contacts. When the main contact is broken, an arc is set up between the arcing horns. The arc moves upward due to the combined action of the hot air currents it produces and the magnetic field.* As the arc rises, it becomes longer until it eventually blows out. Although the arcing horns become pitted and gradually wear out, they can easily be replaced.

Figure 9.4　One pole of a horn-gap disconnecting switch rated 600A, 27kV, 60Hz. (Left) in the open position, (right) in the closed position

Unlike air-break switches, disconnecting switches are unable to interrupt any current at all. They must only be opened and closed when the current is zero. They are isolating switches, enabling us to isolate oil circuit breakers, transformers, transmission lines, and so forth, from a live network. Disconnecting switches is essential to carry out maintenance work and reroute power flow. Disconnecting switches have single-column disconnecting switch, double column disconnecting switch, three column disconnecting switches, V-type disconnecting switches, etc. (Figure 9.5).

a) single column disconnecting switch　　　b) double column disconnecting switch

c) three column disconnecting switch　　　d) V type disconnecting switch

Figure 9.5　Different types of disconnecting switches

A fuse cutout is essentially a fused disconnecting switch. The fuse constitutes the movable arm of the switch. It pivots about one end and the circuit can be opened by pulling on the

# Chapter 9　Switch Devices

other end of the fuse with a hook stick.* Cutouts are relatively inexpensive and are used to protect transformers and small single-phase feeders against overloads. They are designed so that when the fuse blows, it automatically swings down, indicating that a fault has occurred on the line.

## Part 3　Fuse

A fuse refers to an electrical appliance that breaks the circuit by fusing the melt with the heat generated by itself when the current exceeds the specified value.* Fuses are widely used in high and low-voltage distribution systems and control systems as well as in electrical equipment, as a short-circuit and overcurrent protector, it is one of the most commonly used protection devices.

**1. Structure of fuse**

The fuse has a simple structure and consists of an insulating base (or support), contacts, melt, etc. The melt is the main working part of the fuse.

- The melt is equivalent to a special piece of wire connected in series in a circuit. When a short circuit or overload occurs in the circuit, the current is too high and the melt melts due to overheating, thus cutting off the circuit.
- The fuse is generally installed in the fuse housing, which can quickly extinguish the arc.
- The melt is often made into threadiness, grid, or flake, which is generally made of lead-tin alloy, silver-plated copper sheet, zinc, silver, and other metals. It has the characteristics of a low relative melting point, stable characteristics, and easy fusing.

**2. Selection of fuse**

The type of fuse is selected mainly based on the protection characteristics of the load and the size of the short-circuit current.

- Fuses are used for overload and short-circuit protection for small-capacity motors and lighting branches. Hope that the melting coefficient of the fuse is appropriately small, usually choose the RQA series fuse of lead-tin alloy melt.*
- For larger capacity motors and lighting trunk lines, focus on short-circuit protection and breaking capacity, usually using RM10 and RL1 series fuses with high breaking capacity.
- For the case of a large short-circuit current, it is appropriate to use the fuses of RT0 and RT12 series with current limiting effects.

## Part 4　Recloser and Sectionalizers

A recloser is a circuit breaker that opens on a short circuit and automatically recloses after a brief time delay. The delay may range from a fraction of a second to several seconds. The open/close sequence may be repeated two or three times, depending on the internal control setting of the recloser. If the shortcircuit does not clear itself after two or three attempts to reclose the line, the recloser opens the circuit permanently. A repair crew must then locate the fault, remove it, and reset the recloser.

When a main feeder is protected by several fuses spotted over the length of the line, it is often difficult to obtain satisfactory coordination between them, based on fuse-blowing time alone.* Under these circumstances, we resort to a sectionalizer. A sectionalizer is a special circuit breaker that trips depending on the number of times a recloser has tripped further up the line. In other words, a sectionalizer works according to the "instructions" of a recloser.

For example, consider a recloser R and a sectionalizer S protecting an important main feeder (Figure 9.6). If a fault occurs at the point shown, the recloser will automatically open and reclose the circuit, according to the predetermined program. A reorder inside the sectionalizer counts the number of times the recloser has tripped and just before it trips for the last time, the sectionalizer itself trips--but permanently. In so doing, it deprives customers C and D of power but it also isolates the fault. Consequently, when the recloser closes for the last time, it will stay closed and customers A and B will continue to receive service. Unlike reclosers, sectionalizers are not designed to interrupt line currents. Consequently, they must trip during the interval when the line current is zero, which coincides with the time when the recloser itself is open.

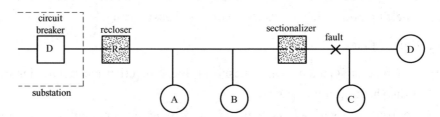

**Figure 9.6** Recloser / sectionalizer protective system

When a fault occurs, the current increases sharply, not only on the faulted line but on all lines that directly or indirectly lead to the shortcircuit. To prevent the overload current from simultaneously tripping all the associated protective devices, we must design the system so that the devices trip selectively.

A well-coordinated system will cause only those devices next to the short-circuit to open, leaving the others intact. To achieve this, the tripping current and tripping time of each device is set to protect the line and associated apparatus, while restricting the outage to the smallest number of customers. The most important protective devices used on MV (Middle Voltage) lines are the following: ① fused cutouts, ② reclosers, and ③ sectionalizers.

# Chapter 9  Switch Devices

## Part 5  The Selection of Circuit Breakers

Two of the factors upon which the proper selection of circuit breakers depends are the current flowing immediately after the fault occurs and the current which the breaker must interrupt.*

The sub-transient current is the initial symmetrical current and does not include the DC component. The inclusion of the DC component results in an rms value of current immediately after the fault, which is higher than the sub-transient current. For oil circuit breakers above 5kV, the sub-transient current multiplied by 1.6 is considered to be the rms value of the current whose disruptive forces the breaker must withstand during the first half cycle after the fault occurs.* This current is called the *momentary current*, and for many years circuit breakers were rated in terms of their momentary current as well as other criteria.

The interrupting rating of a circuit breaker was specified in kilovolt amperes or megavolt amperes. The interrupting kilovolt amperes equal $\sqrt{3}$ times the kilovolts of the bus to which the breaker is connected times the current which the breaker must be capable of interrupting when its contact part.* This current is, of course, lower than the momentary current and depends on the speed of the breaker, such as 8, 5, 3, or 1.5 cycles, which is a measure of the time from the occurrence of the fault to the extinction of the arc.

The current that a breaker must interrupt is usually asymmetrical since it still contains some of the decaying DC components. A schedule of preferred ratings for AC high-voltage oil circuit breakers specifies the interrupting current ratings of breakers in terms of the component of the asymmetrical current which is symmetrical about the zero axes. This current is properly called the *required symmetrical interrupting capability* or simply the *rated symmetrical short-circuit current*. Often the adjective symmetrical is omitted. Selection of circuit breakers may also be made on the basis of total current (DC component included). We shall limit our discussion to a brief treatment of the symmetrical basis of breaker selection.

Breakers are identified by nominal-voltage class, such as 69kV. Among other factors specified are rated continuous current, rated maximum voltage, voltage range factor *K*, and rated short-circuit current at rated maximum kilovolts. *K* determines the range of voltage over which rated short-circuit current times operating voltage is constant.*

## NEW WORDS AND EXPRESSIONS

| | | | |
|---|---|---|---|
| 1. | auto-reclosing | | 自动重合闸 |
| 2. | proving | *n.* | 校对 |
| 3. | mechanical design | | 构造设计 |
| 4. | contact | *n.* | 接触器 |
| 5. | electrical stress | | 电气压力 |
| 6. | nameplate | *n.* | 名牌，标示牌 |
| 7. | tripping coil | | 解扣线圈，跳闸线圈 |

| 8. | deionization | n. | 消电离作用 |
| 9. | triggering action | | 触发作用 |
| 10. | current transformer | | 电流互感器 |
| 11. | preset | vt. | 事先整定 |
| 12. | porcelain bushing | | 陶瓷套管 |
| 13. | actuate | vt. | 开动；促使 |
| 14. | rod | n. | 杆，棒 |
| 15. | courtesy of | | 由…提供，经…允许 |
| 16. | spring | n. | 弹簧 |
| 17. | pull on | | 穿，戴；继续拉 |
| 18. | turbulence | n. | （液体或气体的）紊乱 |
| 19. | swirl | v. | 打漩，盘绕 |
| 20. | air-blast | | 空中爆炸 |
| 21. | supersonic | adj. | 超音速（的） |
| 22. | reservoir | n. | 储液器，气囊 |
| 23. | replenish | v. | 补充 |
| 24. | noise-suppression | | 噪声抑制 |
| 25. | ionize | v. | 电离 |
| 26. | hermetically sealed | | 密封 |
| 27. | arcing horn | | 电弧触角 |
| 28. | pitted | adj. | 有凹痕的 |
| 29. | wear out | | 用坏；磨损 |
| 30. | hook | n. | 钩，吊钩 |
| 31. | stick | n. | 棍，棒 |
| 32. | recloser | n. | 自动开关，自动继电器 |
| 33. | sectionalizer | n. | 分段隔离开关 |
| 34. | repair crew | | 修理队 |
| 35. | fuse | n. | 熔断器 |
| 36. | melt | n. | 熔体 |
| 37. | extinguish | v. | 熄灭；消灭，使破灭 |
| 38. | arc | n. | 弧光，电弧 |
| 39. | threadiness | n. | 像线；线状；丝状 |
| 40. | lead-tin alloy | | 铅锡合金 |
| 41. | melting coefficient | | 熔化系数 |
| 42. | trunk | n. | 树干；驱干 |
| 43. | outage | n. | 断电；临时停止供应（尤指电力） |
| 44. | sub-transient | n. | 次暂态 |
| 45. | asymmetrical | adj. | 不均匀的，不对称的 |
| 46. | adjective | adj. | 从属的，不独立的 |

# PHRASES

**1. depart from: 离开，偏离；从……出发**

Example: Thus, circuit breakers will automatically open a circuit whenever the line current, line voltage, frequency, and so on, departs from a preset limit.

因此，当线电流、线电压、频率等偏离预先整定的限值时，断路器会自动断开电路。

**2. a matter of: 大约，大概**

Example: The most powerful circuit breakers can typically open short-circuit currents of 40kA at a line voltage of 765kV in a matter of 3 to 6 cycles on a 60Hz line.

最有力的断路器典型地可以在 3 到 6 个周期内切断 60Hz、765kV 线路上的 40kA 短路电流。

**3. be equivalent to: 相当于**

Example: The melt is equivalent to a special piece of wire connected in series in a circuit.

熔体相当于串联在电路中的一段特殊的导线。

**4. cut off: 切断；中断**

Example: When a short circuit or overload occurs in the circuit, the current is too high and the melt melts due to overheating, thus cutting off the circuit.

当电路发生短路或过载时，电流过大，熔体因过热而熔化，从而切断电路。

**5. be appropriate to: 合适；适合于**

Example: For the case of a large short-circuit current, it is appropriate to use the fuses of RT0 and RT12 series with current limiting effect.

对于短路电流很大的情况，适合于采用具有限流作用的 RT0 和 RT12 系列熔断器。

**6. a fraction of: 若干分之一，一小部分**

Example: The delay may range from a fraction of a second to several seconds.
延时的范围可能从一秒的若干分之一到几秒。

**7. resort to: 诉诸于，采取**

Example: Under these circumstances, we resort to a sectionalizer.
在这些情况下，我们求助于分段隔离开关。

**8. deprive sb. of sth.: 剥夺某人的某事物，使某人丧失某事物**

Example: In so doing, it deprives customers C and D of power but it also isolates the fault.

这样做，使用户 C 和 D 停电，但是也隔离了故障。

## COMPLICATED SENTENCES

**1. Three movable contacts, actuated simultaneously by an insulated rod, open and close the circuit.**

译文：由一根绝缘杆同时开动的三个移动触点断开和连接电路。

说明：句首的 actuated simultaneously by an insulated rod 是定语从句的缩写，补全应为：which are actuated simultaneously by an insulated rod。

2. When the circuit breaker is closed, the line current for each phase penetrates the tank by way of a porcelain bushing, flows through the first fixed contact, the movable contact, the second fixed contact, and then on out by a second bushing.

译文：当断路器合闸时，每一相的线电流经由陶瓷套管穿入油箱，流经第一个固定触点、移动触点、第二个固定触点，然后经由第二个套管向外。

说明：词组 by way of 表示"经由，通过"。on out 放在一起，表示"向外"。

3. These totally enclosed circuitbreakers, insulated with $SF_6$ gas, are used whenever space is at a premium, such as in down-town substations.

译文：这些用 $SF_6$ 气体隔离的完全密封的断路器，用在进城变电站等空地十分宝贵的场合。

说明：短语 insulated with $SF_6$ gas 是定语从句缩写。补全应为：those are insulated with $SF_6$ gas。

4. Air-break switches are composed of a movable blade that engages a fixed contact; both are mounted on insulating supports.

译文：空气断路开关由一个与固定触点接合的移动闸刀构成；两个（触点）都安装在绝缘支撑上。

5. It pivots about one end and the circuit can be opened by pulling on the other end of the fuse with a hook stick.

译文：它一端为轴转动，用勾杆拉熔断器另外一端可以将电路断开。

说明：词组 pull on 意思为"持续拉"。pivot 表示"在轴上转动"。

6. A fuse refers to an electrical appliance that breaks the circuit by fusing the melt with the heat generated by itself when the current exceeds the specified value.

译文：熔断器是指当电流超过规定值时，以本身产生的热量使熔体熔断，断开电路的一种电器。

说明：句子中包括 that 引导的定语从句和 when 引导的时间状语从句，that 引导的定语从句用来修饰句子的主语"fuse"，when 引导的时间状语从句表示熔断器熔体熔断这一动作发生在流过熔断器的电流超过规定值时。"with the heat generated by itself"中过去分词"generated"作后置定语来修饰名词"heat"。

7. Hope that the melting coefficient of the fuse is appropriately small, usually choose the RQA series fuse of lead-tin alloy melt.

译文：希望熔体的熔化系数适当小些，通常选用铅锡合金熔体的 RQA 系列熔断器。

说明：句子开头为祈使句，祈使句一般不出现主语。然后是 that 作连词引导的宾语从句，此时 that 只起到连接主、从句的作用，它本身无意义，在口语或非正式文体中常省略。逗号后面的语句同样也为祈使句。

# Chapter 9  Switch Devices

**8. When a main feeder is protected by several fuses spotted over the length of the line, it is often difficult to obtain satisfactory coordination between them, based on fuse-blowing time alone.**

译文：当一条主馈线由沿着线路长度方向斑驳分布的几个熔断器保护时，仅仅基于保险丝熔断时间，常常难以实现它们之间令人满意的配合。

说明：由 based on 引导的状语短语，表示前面陈述成立的条件，在翻译时这样的状语应该提前。

**9. Two of the factors upon which the proper selection of circuit breakers depends are the current flowing immediately after the fault occurs and the current which the breaker must interrupt.**

译文：合理选择断路器取决于两个因素，即故障发生后立刻流过的电流和断路器必须关断的电流。

说明：定语从句 upon which the proper selection of circuit breakers depends 中谓语短语的介词提前到了引导词以前。正常语序为：which the proper selection of circuit breakers depends upon。

**10. For oil circuit breakers above 5kV, the sub-transient current multiplied by 1.6 is considered to be the rms value of the current whose disruptive forces the breaker must withstand during the first half cycle after the fault occurs.**

译文：对于5kV以上的油断路器来说，将次暂态电流乘以1.6，被看作是故障发生后的前半个周期内断路器必须经受其破坏力的那个电流的有效值。

说明：这个句子翻译起来比较拗口，原因就是 whose 引导了一个较长的定语从句。为了理解和翻译方便，可以将该句子拆解为：

For oil circuit breakers above 5kV the sub-transient current multiplied by 1.6 is considered to be the rms value of that current; the breaker must withstand its disruptive forces during the first half cycle after the fault occurs.

**11. The interrupting kilovolt amperes equal $\sqrt{3}$ times the kilovolts of the bus to which the breaker is connected times the current which the breaker must be capable of interrupting when its contact part.**

译文：关断功率千伏安等于$\sqrt{3}$乘以断路器连接的母线电压，以千伏计，再乘以断路器在接点分开以后可以关断的电流。

说明：这个句子里，kilovoltampere 意为千伏安，在句中表示以千伏安计算的功率。这个句子的主结构是 kilovoltampere equals $\sqrt{3}$ times kilovolts times current。

**12. K determines the range of voltage over which rated short-circuit current times operating voltage is constant.**

译文：K决定电压的范围，在这个范围内，额定短路电流和工作电压的积为常数。

## ABBREVIATIONS (ABBR.)

1. CB　　　circuit breaker　　　　断路器
2. OCB　　oil circuit breaker　　　油断路器，油开关
3. MV　　　middle voltage　　　　中压

## SUMMARY OF GLOSSARY

1. 断路器　circuit breakers
   　　Oil Circuit Breakers　　　　　　油断路器
   　　Air-Blast Circuit Breakers　　　空中爆炸断路器
   　　$SF_6$ Circuit Breakers　　　　　六氟化硫断路器
   　　Vacuum Circuit Breakers　　　真空断路器
2. 开关　switches
   　　air-break switches　　　　　　空气断路开关
   　　disconnecting switches　　　　隔离开关，断路开关
   　　isolating switches　　　　　　隔离开关
   　　fuse cutout　　　　　　　　　保险器，熔丝断路器
3. 保护设备　protection devices
   　　circuit breaker　　　　　　　　断路器
   　　switch　　　　　　　　　　　开关
   　　fuse　　　　　　　　　　　　熔断器
   　　recloser　　　　　　　　　　重合闸设备，自动开关
   　　sectionalizer　　　　　　　　分段隔离开关

# EXERCISES

**1. Translate the following words or expressions into Chinese.**

(1) nameplate　　　　　(2) tripping coil　　　　(3) contingency

(4) air-blast　　　　　　(5) recloser　　　　　　(6) porcelain bushing

(7) outage

**2. Translate the following words or expressions into English.**

(1) 接触器　　　　　　(2) 完好无缺的　　　　　(3) 分段隔离开关

(4) 密封　　　　　　　(5) 噪声抑制

**3. Fill in the blanks with proper words or expressions.**

(1) The nameplate on a circuit breaker usually indicates (a) the maximum _____ _____ current it can carry, (b) the maximum _____ current, (c) the maximum line _____, and (d) the interrupting _____ in cycles.

(2) The most important protective devices used on MV (Middle Voltage) lines are the

following: ① _____, ② _____, and ③ _____.

# Word-Building (9)　ab-, a-　否定 3

**1. ab-[ 前缀 ]，表示"偏离，脱离或离开"之义，可译为"不，无，反，异"**

| | | | |
|---|---|---|---|
| end | abend | n. | 异常结束，异常终止 |
| joint | abjoint | adj.n. | 分离（的），分节（的） |
| migration | abmigration | n. | 反常迁徙 |
| normal | abnormal | adj. | 反常的，变态的 |
| react | abreact | vt. | 使消散，发泄 |

**2. a-[ 前缀 ]，表示"偏离，脱离或离开"之义（用于 s 前）**

| | | | |
|---|---|---|---|
| synchronous | asynchronous | adj. | 异步的；不同时的 |
| seasonal | aseasonal | adj. | 无季节性的 |
| seismic | aseismic | adj. | 无地震的，耐震的 |
| semantide | asemantide | n. | 无信息分子 |
| septic | aseptic | adj. | 防腐的，无菌的 |
| sexual | asexual | adj. | 无性的，无性生殖的 |
| symmetrical | asymmetrical | adj. | 不均匀的，不对称的 |

# Unit 3

## Review

### I. Summary of Glossaries: Electric Machines and Breakers

1. 电机　electric machine
   - direct-current machine　　　　　　　　　　　直流电机
   - alternating-current machine　　　　　　　　　交流电机
   - synchronous machine　　　　　　　　　　　　同步电机
   - asynchronous machine　　　　　　　　　　　　异步电机
   - induction machine　　　　　　　　　　　　　　感应电机

2. 电机的部件　parts in electric machines
   - rotor　　　　　　　　　　　　　　　　　　　　转子
   - stator　　　　　　　　　　　　　　　　　　　　定子
   - armature　　　　　　　　　　　　　　　　　　电枢

3. 电机的负载　loads of motors
   - full-load　　　　　　　　　　　　　　　　　　满载
   - no-load　　　　　　　　　　　　　　　　　　　空载

4. 保护设备　protection devices
   - circuit breaker　　　　　　　　　　　　　　　断路器
   - switch　　　　　　　　　　　　　　　　　　　开关
   - recloser　　　　　　　　　　　　　　　　　　重合闸设备，自动开关
   - sectionalizer　　　　　　　　　　　　　　　　分段隔离开关
   - relay　　　　　　　　　　　　　　　　　　　　继电器

5. 感应
   - induce　　　　　　　　　　vt.　　　　　　　　感应，引起
   - induct　　　　　　　　　　vt　　　　　　　　 感应，感生

6. 转动
   - turn　　　　　　　　　　　vt.n.　　　　　　　（使）转动，翻转
   - rotate　　　　　　　　　　vt. rotation n.　　 （使）旋转
   - revolve　　　　　　　　　 vt.　　　　　　　　（使）旋转
   - revolving　　　　　　　　 v.　　　　　　　　 旋转
   - revolution　　　　　　　　n.　　　　　　　　 旋转
   - running　　　　　　　　　n.　　　　　　　　 转动，运转

### II. Abbreviations (Abbr.)

1. RMS　　　　　　root mean square　　　　　　　　　　均方根

| | | | |
|---|---|---|---|
| 2. | MMF | magnetomotive force | 磁动势，磁通势 |
| 3. | EMF | electromotive force | 电动势 |
| 4. | CB | circuit breaker | 断路器 |

# Special Topic(3)　Powers of Ten in Engineering Notation

Specific powers of ten in engineering notation have been assigned prefixes and symbols, as appearing in Table 1. They permit easy recognition of the power of ten and an improved channel of communication between technologists.

Table 1　prefixes and symbols of specific powers of ten

| Multiplication Factors | SI Prefix | SI Symbol | |
|---|---|---|---|
| $1\,000\,000\,000\,000 = 10^{12}$ | tera | T | 太（拉），万亿 |
| $1\,000\,000\,000 = 10^{9}$ | giga | G | 吉（咖），十亿，千兆 |
| $1\,000\,000 = 10^{6}$ | mega | M | 兆，百万 |
| $1\,000 = 10^{3}$ | kilo | k | 千 |
| $0.001 = 10^{-3}$ | milli | m | 毫，千分之一 |
| $0.000\,001 = 10^{-6}$ | micro | $\mu$ | 微，百万分之一 |
| $0.000\,000\,001 = 10^{-9}$ | nano | n | 纳，十亿分之一 |
| $0.000\,000\,000\,001 = 10^{-12}$ | pico | p | 皮，兆分之一 |

**Example:**

a. $1,000,000 \text{ohms} = 1 \times 10^{6} \text{ohms} = 1 \text{megohm}$

b. $41,200 \text{meters} = 41.2 \times 10^{3} \text{meters} = 100 \text{kilometers} = 100 \text{km}$

c. $0.0001 \text{second} = 0.1 \times 10^{-3} \text{second} = 0.1 \text{millisecond} = 100 \mu \text{s}$

d. $0.000001 \text{farad} = 1 \times 10^{-6} \text{farad} = 1 \text{microfarad} (\mu F)$

e. A terawatt is equal to a trillion watts, that is $10^{12}$W.

## Word-Building (F)　构词方法——数量级

### 1. meg(a)- 表示：兆，百万（量级）

| byte | megabyte | n. | MB | 兆字节，百万字节 |
|---|---|---|---|---|
| bit | megabit | n. | Mb | 兆位，百万位 |
| hertz | megahertz | n. | MHz | 兆赫兹 |
| farad | megafarad | n. | MF | 兆法（拉） |
| joule | megajoule | n. | MJ | 兆焦（耳） |
| ohm | megohm | n. | MOhm | 兆欧 |
| voltampere | megavoltampere | n. | MVA | 兆伏安 |
| watt | megawatt | n. | MW | 兆瓦 |

**2. kilo-**　表示：千，千倍（量级）

| volt | kilovolt | n. | kV | 千伏 |
| voltampere | kilovoltampere | n. | kVA | 千伏安 |
| watt | kilowatt | n. | kW | 千瓦 |
| gram | kilogram | n. | kg | 千克，公斤 |
| meter | kilometer | n. | km | 千米，公里 |
| litre | kilolitre | n. | kl | 千升（容量单位） |

# Knowledge & Skills (3)　　Omitting of Nouns

**1. 宾语的省略**

（1）当多个动词或动词词组带有共同的宾语时，往往会进行适当的处理，共同的宾语只出现一次，其余的可以省略。

例 1：Synchronous generators can generate or absorb reactive power depending on the excitation.

由于励磁的不同，同步发电机可以发出或者吸收无功功率。

说明：动词 generate 和 absorb 具有共同的宾语 reactive power，该宾语只出现了一次，generate 后面的宾语省略了。句子补全应为：Synchronous generators can generate reactive power or absorb reactive power depending on the excitation.

（2）当多个介词带有共同的宾语时，往往会进行适当的处理，共同的宾语只出现一次，其余的可以省略。

例 1：They are "distributed" because they are placed at or near the point of energy consumption.

它们是"分布式的"，是因为它们位于或接近能量消耗的地点。

说明：at 和 near 具有相同的宾语 the point。在介词 at 后面省略了宾语，补全应为 placed at the point or near the point。

（3）当多个"动词+介词"形式的词组带有共同的宾语时，往往会进行适当的处理，共同的宾语只出现一次，其余的可以省略。

例 1：Depending on how the connection is made, the secondary voltage may add to, or subtract from the primary voltage.

基于连接方式，二次侧电压可以叠加到一次侧电压上，或者从一次电压中减掉。

说明：动词词组 add to 和 subtract from 具有共同的宾语 the primary voltage，该宾语只出现了一次，add to 后面的宾语省略了。句子补全应为：... the secondary voltage may add to the primary voltage or subtract from the primary voltage。

例 2：The electromagnetic environment for disturbances originating in or conducted through the power system is equivalent to the voltage quality.

源于或通过电力系统传导的扰动，其电磁环境等同于电压质量。

说明：originating in 和 conducted through 具有共同的宾语 the power system。在 originating in 后面省略了宾语，补全应为：... originating in the power system or conducted through the power system。

例 3：The rapid and frequent voltage changes, especially flickers, caused by these devices can be perceived by and be objectionable to nearby customers and often result in complaints to the power companies.

这些设备引起的快速而频繁的电压变化，尤其是闪变，能被附近用户感知而且比较令人讨厌，常常导致对电力公司的抱怨。

说明：be perceived by 和 be objectionable to 具有相同的宾语 nearby customers 中，在 by 的后面省略了宾语。补全应为：be perceived by nearby customers and be objectionable to nearby customers。当然，在翻译时也可以做适当的省略。

**2. 名词短语中主体名词的省略**

（1）同时列举多个类似概念时，在不引起混淆的情况下，这些类似概念中所包含的相同词语往往省略。

例 1：The ratio of primary to secondary voltage is also called the turn ratio.

一次电压和二次电压之比也称为变比。

说明：在 primary to secondary voltage 中省略了 primary 后面的 voltage，补全应为：The ratio of primary voltage to secondary voltage。

例 2：Because of their common origin, the temporary and switching overvoltages occur together and their combined effect has to be taken into account in the design of HV systems insulation.

由于有共同的起因，临时过电压和操作过电压同时发生，其综合影响在高压系统绝缘设计中必须引起重视。

说明：在 temporary and switching overvoltages 中省略了 temporary 后面的 overvoltages，补全应为：temporary overvoltages and switching overvoltages

例 3：Discharges also occur between positive and negative charges within the cloud, rather than between the base of the cloud and the ground.

放电也会出现在云层内部的正负电荷之间，多于出现在云层底部和大地之间。

说明：在 positive and negative charges 中省略了 positive 后面的 charges，补全应为：positive charges and negative charges。

例 4：The effective use of grid-connected distributed energy resources can also require power electronic interfaces and communication and control devices for efficient dispatch and operation of generating units.

联网分布式能源的有效利用，也需要电力电子接口、通信设备和控制设备用于发电机组的调度和运行。

说明：在 communication and control devices 中省略了 communication 后面的 devices，补全应为：communication devices and control devices

例 5：Some of the most common devices are induction and synchronous motors.

最常见的一些设备是感应电机和同步电机。

说明：在 induction and synchronous motors 中 induction 的后面省略了 motors。

（2）当一个名词作为多个名词短语的主体时，在不引起混淆的情况下，可能会省略其中一部分。

例1：The frequency of the induced alternating voltages and of the resulting currents that flow in the stator windings when a load is connected depends on the speed of the rotor.

感应出的交流电压的频率和当接有负荷时流过定子的电流的频率，取决于转子的速度。

说明：在翻译时，可以认为连词 and 连接了两个并列的主体，共同作为整个句子的主语，共用一个谓语动词 depend。这两个主体分别是 the frequency of the induced alternating voltages 和 the frequency of the resulting currents，其中第二个短语的中心词 frequency 省略了。句子补全应为：The frequency of the induced alternating voltages and the frequency of the resulting currents that flow in the stator windings when a load is connected depend on the speed of the rotor. 注意：在翻译时可以这样理解，由原句的谓语动词 depends 的第三人称单数形式可知，其实在这个英文句子里只有一个 frequency 作为句子主语，而它是由两个 of 引导的短语共同修饰的，这也正是中英文表达习惯的差别。

# Chapter 10
# Instruments and Transducers

## Part 1  Importance of Instrumentation

A modern electric power system is an assembly of many components each of which influences the behavior of every other part. Proper functioning of the system as a whole makes it necessary to monitor conditions existing at many different points in the system in order to ensure optimum operation.*

The concern of the customers is primarily that the frequency and voltage of the supply are held within certain rather narrow limits. Since the frequency of the system may be quite different at different points, it is necessary to make continuous observation of the voltage at certain key points on the system in order to provide acceptable service.

Instrumentation is necessary to permit the billing of customers for energy used. Many interconnections exist between different power systems. Instruments must be provided at interchange points to permit billing for energy transferred from one system to another.* The continuous monitoring of energy transfer is necessary to ensure that interchanged power is within the limits of contract agreements.

The continuous measurement of conditions on major pieces of equipment is necessary to avoid damage due to overload. As load increases from month to month, points at which additional capacity of the equipment is required may be recognized and provision made for the installation of additional equipment.* Thus instrumentation serves as a guide for future construction in a growing power system.

Occasionally, under emergency conditions, a system operator observes that his system load exceeds the ability of the available generating and transmission equipment. He is then faced with the problem of load shedding. It is then necessary to drop selected loads where service interruption is least objectionable. In such an event, he relies on the instruments which provide information relative to system-operation conditions.

Instruments may sound alarms as warnings of conditions requiring action to avoid damage to equipment operating beyond its design limitation.* In the event of extreme conditions such as power-system faults, defective equipment is switched out of service automatically. Instruments that continuously monitor current, voltage, and other quantities must be able to identify the faulted equipment and to bring about the operation of the circuit breakers which remove it from service, while leaving in service all other equipment on the operating system.*

The many different electrical devices on a power system and those owned by the customers are designed for operation within certain specified ranges. Operation outside these designed limits is undesirable, as it may result in inefficiency of operation, excessive deterioration, or (in extreme cases) the destruction of the device. Attention to the conditions under which equipment is operating may indicate corrective action that must be taken.

## Part 2  Effects without Instrumentation

Abnormal current, voltage, and speed will occur without instrumentation.

Overcurrent on all electrical devices is undesirable, as it produces excessive temperatures, inefficient operation, and reduced service life. Overcurrent in residential circuits may bring about disconnection of the circuit by fuse or breaker action. Overcurrent in motors may damage insulation, with possible early insulation failure.

Undervoltage considerably reduces the efficiency of incandescent lamps and may result in non-operation of fluorescent lamps. Undervoltage of the power supply to motors may result in excessive currents in the motors, with possible damage to windings.

Overvoltage increases the light output of lamps but in many instances seriously shortens useful life. Overvoltage applied to motors and transformers may result in excessive losses within the iron, with possible damage to the iron or the adjacent winding insulation.

Overspeed of rotating machines may result in structural damage to rotating parts. The overspeed of the customer's production equipment may result in an inferior quality of the product.

An out-of-step condition existing between two generators or between a generator and a synchronous motor results in an interruption of useful power transfer between the two machines. An out-of-step condition should be recognized promptly and the machines separated from each other. They may then be resynchronized and brought back into service.

Instruments of many different types must be installed at many locations on a power system and the premises of many customers. With such instruments, conditions existing in the system may be continuously monitored.

## Part 3  Instruments for System Monitoring

The instruments used for system monitoring include the following:

| | |
|---|---|
| voltmeters | synchroscopes |
| ammeters | relays |
| watt meters | varmeters |
| watt-hour meters | automatic oscillographs |
| frequency meters | |

# Chapter 10  Instruments and Transducers

The meters may be indicating or recording. The instruments may in turn control automatic equipment for regulating voltage, frequency, power output, or power interchange on tie lines. In many modern systems, instrument readings are transmitted from many points on the operating system to control centers. At these control centers the reading of instruments from the remote points may be displayed as an aid to the system dispatcher. In some instances, data from selected instruments are fed into a computer which determines plant loading for the most economic system operation. The computer then transmits signals to the various generators to load each machine to the predetermined value.

The transmission of instrument readings from point to point on a power system may be accomplished by several different methods. Leased or private telephone lines may provide circuits for information transmission. Channels of communication may be provided by carrier current systems using the power conductors as the medium of transmission. In other instances, microwave circuits may be utilized. The system used for the transmission of instrument readings may also be used for voice transmission and supervisory control.

## Part 4  Transducers

Currents and voltages of the protected power equipment are converted by current and voltage transformers to low levels for meter measurement or relay operation. These reduced levels are necessary for two reasons: ① the lower level input to the instruments ensures that the physical hardware used to construct the instruments will be quite small and thus less expensive; ② the personnel who work with the instruments will be working in a safe environment.

In principle, these transducers are no different from the power transformers. However, the use made of these transformers is rather specialized. For example, a current transformer reproduces in its secondary winding a current that duplicates the primary current waveform as faithfully as possible.* It performs this function quite well. Similar considerations hold for a voltage transformer. The amount of power delivered by these transformers is rather modest since the load connected to them consists only of relays and meters that may be in use at a given time. The load on current transformers (CTs) and voltage transformers (VTs) is commonly known as their *burden*. The term burden usually describes the impedance connected to the transformer's secondary winding but may specify the volt-apmperes delivered to the load. For example, a transformer delivering 5 amperes to a resistive burden of 0.1 ohm may also be said to have a burden of 2.5 voltamperes at 5 amperes.

### 1. Current Transformers

There are two types of current transformers found in practice. Certain power equipment (such as power transformers, reactors, and oil circuit breakers) is of the dead-tank type, having a grounded metal tank in which the power equipment is contained in an insulating medium (usually oil) and a bushing through which a terminal of the equipment is brought out.*

Current transformers are built within this bushing and are known as bushing CTs. Where such a dead-tank system is not available, for example at an EHV switching station where live-tank circuit breakers are in use, free-standing current transformers are used.*

Current transformers have ratio errors which for some types can be calculated and for other types must be determined by test. The error can be quite high if the impedance burden is too large, but with proper selection of the CT with respect to the burden the error can be maintained at an acceptable value.

**2. Voltage Transformers**

Two types of voltage transformers are commonly found in applications. For certain low-voltage (about 12kV or lower) applications transformers with a primary winding at the system voltage and secondary windings at 67V (line-to-neutral voltage) and 116V (line-to-line voltage) are an industry standard. This type of voltage transformer is quite similar to a multi-winding power transformer and becomes expensive at higher system voltages. For voltages at HV and EHV levels, a capacitance potential-divider circuit is used. In such a coupling-capacitor voltage transformer (CVT) the tapped voltage is further reduced to metering or relaying voltage level by a transformer.

Voltage transformers are generally far more accurate than current transformers, and their ratio and phase-angle errors are generally neglected. On the other hand, it is often necessary to pay attention to the transient response of the CVTs under fault conditions, as errors under these conditions are possible.

## NEW WORDS AND EXPRESSIONS

| 1. | instrument | n. | 仪表，仪器 |
| 2. | transducer | n. | 传感器，变换器 |
| 3. | bill | v. | 把…列成表，给…开账单 |
| 4. | load shedding | | 甩负荷 |
| 5. | objectionable | adj. | 引起反对的，讨厌的 |
| 6. | deterioration | n. | 变坏，退化 |
| 7. | incandescent lamp | | 白炽灯，白热灯 |
| 8. | out-of-step | | 步调不一致，不同步的 |
| 9. | premise | n. | 房产，房屋及地基等 |
| 10. | meter | n. | 计，表，仪表 |
| 11. | synchroscope | n. | 同步示波器，同步检定器 |
| 12. | oscillograph | n. | 示波器，录波器 |
| 13. | tie line | | 联络线 |
| 14. | reading | n. | 读数，指示数 |
| 15. | dispatcher | n. | 调度员 |
| 16. | predetermined | adj. | 预定的 |
| 17. | leased | adj. | 租用的 |
| 18. | carrier | n. | 载波（信号） |
| 19. | supervisory control | | 监控 |

# Chapter 10  Instruments and Transducers

| 20. | voltage transformer | | 电压互感器 |
|---|---|---|---|
| 21. | modest | *adj.* | 不多的，不大的 |
| 22. | burden | *n.* | 负载 |
| 23. | dead-tank | | 固定箱体 |
| 24. | ratio error | | 比率误差 |
| 25. | potential-divider | | 分压器 |
| 26. | transient response | | 瞬时反应，暂态反应 |

## PHRASES

**1. be faced with: 面临**

Example: He is then faced with the problem of load shedding.
于是他面临甩负荷的问题。

**2. bring back into: 使恢复**

Example: They may then be resynchronized and brought back into service.
随后它们可以重新同步并恢复运行。

**3. with respect to: 至于，关于，对应于**

Example: The error can be quite high if the impedance burden is too large, but with proper selection of the CT with respect to the burden the error can be maintained at an acceptable value.
如果负载阻抗太大，误差会相当大，但是根据负载适当选择电流互感器可以将该误差维持在可以接受的水平。

## COMPLICATED SENTENCES

**1. Proper functioning of the system as a whole makes it necessary to monitor conditions existing at many different points in the system in order to ensure optimum operation.**

译文：为了使系统作为一个整体正常运行，就有必要监测系统中很多不同地点的情况。

说明：词组 as a whole 表示"整体上"，也可以理解为"作为一个整体"。表示"存在"的 existing 不必翻译。一般来说，表示目的、原因等的状语短语在翻译时都应提前。

**2. Instruments must be provided at interchange points to permit billing for energy transferred from one system to another.**

译文：必须在（能量）交换点配备仪器，以便对从一个系统输送到其他系统的能量进行计费。

说明：英文文献惯用被动语态，而中文表述习惯常用主动语态，在翻译时应考虑这一点。另外，关于某些动词的翻译可以灵活一些，以符合中文习惯。

**3. As load increases from month to month, points at which additional capacity of the equipment is required may be recognized and provision made for the installation of additional equipment.**

译文：随着负荷的逐月增长，（人们）可能认识到那些需要额外容量的地方，并且为额外设备的安装提供条件。

说明：逗号后面的句子是由 and 连接的两个并列关系的句子，主语分别是 points 和 provision，具有共同的谓语结构（能愿动词 may 为谓语主体，后面是被动语态形式的具体内容），其中第二个句子（provision 为主语）的谓语部分省略了 may be，第一句子的主语 points 带有一个由介词＋关系代词构成的 at which 引导的定语从句。为了便于理解，整个句子可以拆解为：

As load increases from month to month, points may be recognized; at these points additional capacity of equipment is required; and provision may be made for the installation of additional equipment.

**4. Instruments may sound alarms as warnings of conditions requiring action to avoid damage to equipment operating beyond its design limitation.**

译文：仪器可以使警报器发声，作为需要采取行动，避免运行于设计极限之外的设备损害。

说明：这里 sound 为谓语动词，表示"使…发声"。alarms 是指"警报器"。为了便于理解，整个句子可以拆解为：

Instruments may sound alarms as warnings of conditions; the conditions require action to avoid damage to equipment; the equipment is operating beyond its design limitation.

**5. Instruments that continuously monitor current, voltage, and other quantities must be able to identify the faulted equipment and to bring about the operation of the circuit breakers which remove it from service, while leaving in service all other equipment on the operating system.**

译文：持续监测电流、电压和其他量的仪表必须能够识别故障设备并触发断路器动作，将其从供电电路切除，而使运行系统中的所有其他设备保持供电。

说明：句中的主语 instruments 带有一个由 that 引导的很长的定语从句，而定语从句中的 circuit breakers 带有一个 which 引导的定语从句。Which 引导的定语从句本身又是一个简化了的复合句，其中第二部分用的是分词短语的形式。词组 bring about 表示"导致，使发生"。为了便于理解，这个句子可以拆解为：

Instruments continuously monitor current, voltage, and other quantities; they must be able to identify the faulted equipment and be able to bring about the operation of the circuit breakers; the circuit breakers remove the faulted equipment from service, while the circuit breakers leave all other equipment on the operating system in service.

**6. A current transformer reproduces in its secondary winding a current that duplicates the primary current waveform as faithfully as possible.**

译文：电流互感器在其二次侧产生一个尽量如实再现一次电流波形的电流。

# Chapter 10　Instruments and Transducers

说明：that 引导的从句是 it 的同位语，在翻译时可以用同位语从句的内容替代 it。

**7. Certain power equipment (such as power transformers) is of the dead-tank type, having a grounded metal tank in which the power equipment is contained in an insulating medium (usually oil) and a bushing through which a terminal of the equipment is brought out.**

译文：某些电力设备（例如电力变压器）是固定箱体型的，带有一个接地的金属箱，电力设备在里面被绝缘介质（通常为油）包围，还有一个导管，电力设备的一端通过它伸出来。

说明：词组 be of ... type 表示"是…类型"。词组 bring out 在此表示"伸出，引出"。句中的 in which 和 through which 都是以介词 + 连词的形式构成的定语从句引导词，其所引导的定语从句分别修饰 metal tank 和 bushing，而这两个被修饰的主体并列作为 having 的宾语。为了便于理解，这个句子可以拆解为：

Certain power equipment (such as power transformers) is of the dead-tank type; the dead-tank type has a grounded metal tank; in a metal tank, the power equipment is contained in an insulating medium (usually oil); the dead-tank type also has a bushing; a terminal of the equipment is brought out through the bushing.

**8. Where such a dead-tank system is not available, for example at an EHV switching station where live-tank circuit breakers are in use, free-standing current transformers are used.**

译文：在这种固定箱体系统不能用的场合，例如在使用活动箱体断路器的超高压开关站，则使用无支撑（自立）的电流互感器。

说明：第一个 where 引导的从句是整个句子的状语，表示场景。第二个 where 引导的从句是定语从句，修饰 station。

## ABBREVIATIONS (ABBR.)

| | | | |
|---|---|---|---|
| 1. | CT | current transformer | 电流互感器 |
| 2. | PT/VT | potential transformer/voltage transformer | 电压互感器 |
| 3. | EHV | extra-high voltage | 极高压，超高压 |
| 4. | CVT | coupling-capacitor voltage transformer | 耦合电容电压互感器 |

## SUMMARY OF GLOSSARY

1. 变换器　**transformers**
   current transformer　　　　　　　　电流互感器
   potential/voltage transformer　　　　电压互感器
   power transformer　　　　　　　　电力变压器
2. 箱体　**tanks**
   dead-tank　　　　　　　　　　　　固定箱体
   live-tank　　　　　　　　　　　　活动箱体
3. 两种电流互感器　**two types of current transformers**
   bushing CT　　　　　　　　　　　套管 CT

    free-standing CT       自立（无支撑）CT
  4. 误差 errors of transformers
    ratio error          比率误差
    phase-angle error       相角误差

# EXERCISES

**1. Translate the following words or expressions into Chinese.**

(1) instrument    (2) monitor    (3) load shedding
(4) inefficiency   (5) out-of-step   (6) synchroscope
(7) carrier line   (8) supervisory control  (9) burden

**2. Translate the following words or expressions into English.**

(1) 读数    (2) 传感器    (3) 示波器
(4) 比率误差  (5) 电压互感器  (6) 暂态反应

**3. Fill in the blanks with proper words or expressions.**

(1) Instruments may sound alarms as warnings of conditions requiring action to avoid _____ to equipment operating beyond its _____.

(2) Instruments that continuously monitor current, voltage, and other quantities must be able to identify the _____ equipment and to bring about the operation of the _____ which remove it from service, while leaving in service all _____ equipment on the operating system.

(3) A current transformer reproduces in its _____ winding a current that _____ the _____ current waveform as faithfully as possible.

# Word-Building (10)  -meter, over-, under 仪表，合成词

**1. -meter 后缀，表示"仪表"，可译为"…计，…表"**

| | | | |
|---|---|---|---|
| ampere | ammeter | n. | 安培计，电流表 |
| volt | voltmeter | n. | 伏特计，电压表 |
| watt | wattmeter | n. | 瓦特计，功率表 |
| var | varmeter | n. | 乏尔计，无功表 |
| micro- | micrometer | n. | 测微计，千分尺 |
| magneto- | magnetometer | n. | 磁力计 |
| thermal | thermometer | n. | 温度计，体温计 |
| pieze | piezometer | n. | 压力计，压强计 |
| tone, tono- | tonometer | n. | 音调计，气压计，张力计 |
| speed | speed meter | n. | 速度计 |

# Chapter 10　Instruments and Transducers

**2. 合成词，over 在前，①表示"过，超"；②表示"在…上面，优越"**

| | | | |
|---|---|---|---|
| load | overload | n. | 超载，负荷过多 |
| voltage | overvoltage | n. | 过电压 |
| current | overcurrent | n.v. | 过电流 |
| charge | overcharge | n.v. | 超载，过度充电 |
| frequency | overfrequency | n. | 过频率，超过额定频率 |
| speed | overspeed | n. | 超速 |
| head | overhead | ad. | 在头上（的），高架（的），在空中 |

**3. 合成词，under 在前，①表示"欠，低"；②表示"在…下面"**

| | | | |
|---|---|---|---|
| load | underload | n.vi. | 欠载，部分载荷，低负载 |
| current | undercurrent | n. | 欠电流，潜流，暗流 |
| voltage | undervoltage | n. | 欠电压，低电压 |
| charge | undercharge | n.vt. | 充电不足，填充不足 |
| excite | under excite | vt. | 欠励磁，励磁不足 |
| frequency | underfrequency | n. | 低于额定频率 |
| power | underpower | n. | 低功率，功率[马力]不足 |
| ground | underground | ad. | 地下（的），秘密（的） |
| water | underwater | ad. | 在水下（的），在水中（的） |
| utilized | underutilized | adj. | 未充分利用的 |

# Chapter 11
# Power System Relay Protection

## Part 1   Relay Type

Relays are the logic elements of a protection system that sense the fault and cause the circuit-breaker trip circuits to be energized and the breakers to open their contacts.* The output state of the relay will be, with its contacts closed, called the trip, or with its contacts open, called block or block to trip. The common types are shown in SUMMARY OF GLOSSARY.

### 1. The working principle of electromagnetic current relays

As shown in Figure 11.1, when the coil of the electromagnet passes the current I, magnetic flux will be generated on the magnetic furnace composed of the electromagnet, movable armature, and air gap. The magnetic flux generates an electromagnetic force in the air gap, which makes the movable armature have the tendency to attract the electromagnet.

When $I < I$op (action current), the armature does not close, the contacts remain open, and the external circuit is not connected. When $I \geq I$op, the armature closes, the contact closes, and the external circuit is connected.

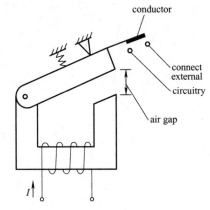

**Figure 11.1**   Structure of electromagnetic relay

Therefore, the current relay can be applied to measure whether the input current meets the condition of $I \geq I$op, which belongs to the mode of excessive action, that is, the action is greater than or equal to the set value.

### 2. Practical application of relay

There are many kinds of electromagnetic relays, which can be divided into three types, namely voltage relay, current relay, and intermediate relay. These are commonly used relays. The following Table 11.1 is the classification and use of universal control electromagnetic relays:

# Chapter 11 Power System Relay Protection

Table 11.1 Classification and use of relays

| Type | Action characteristic | Main application |
| --- | --- | --- |
| Voltage relay | Action when the voltage of the parallel coil reaches the specified action value | For overvoltage protection and voltage loss protection |
| Current relay | Action when the current of the series coil reaches the specified action value | For overload protection or field control loss of excitation protection of DC motor |
| Intermediate relay | A voltage relay in nature, but there are many contacts | Commonly used for amplification and multiplex transmission of switching signals |
| Time relay | When the touch signal of the coil is shifted, the output action signal has a certain delay | For time-related and production process control |
| Thermal relay | The overload current flows through the bimetallic sheet to bend it and push the control mechanism to move | For overload protection and phase loss protection of AC motor |
| Temperature relay | Action when the outside temperature reaches the specified value | For temperature control and motor overheat protection control |
| Speed relay | Action when the rotation rate reaches the specified value | For energy consumption braking and reverse connection braking of motor |

## Part 2  Protection Performance and Protection Area

The relay protection acting on tripping shall technically meet the following requirements:

(1) Speed of operation. A relay used for protection should make its decision as quickly as possible, consistent with other requirements placed upon it.*

(2) Dependability or Reliability. The relay should operate consistently for all the faults for which it is designed to operate and should refrain from operating for any other system condition.*

(3) Selectivity or Discrimination. The smallest possible portion of a system should be isolated following a fault.

(4) Sensitivity. This is the level of magnitude of fault current at which operation occurs.

In general, the reliability requirements of a relay are in conflict with its speed requirement, sensitivity and reliability are a bunch of contradictions, and a compromise must be made in designing the protection system so as to obtain a reasonable measure of these attributes.

(5) Protection zone. The concept of zones helps define the reliability requirements for different protection systems. Each main protection scheme protects a defined area or zone of the power system.

(6) Primary protection. The protection systems considered those are primarily responsible for the removal of the faults as soon as possible while de-energizing as little of the system as required are known as the primary protection systems.

(7) Back-up protection. A completely separate arrangement that operates to remove the faulty part should be the main protection that fails to operate*, including local backup protection and remote backup protection.

# Part 3  Composition of Protection System

Speedy elimination of a fault by the protection system requires the correct operation of a number of subsystems of the protection system. The job of each of these subsystems can best be understood by describing the events that take place from the time of occurrence of a fault to its eventual elimination from the power system.*

The protection system can be subdivided into three subsystems:

(1) Transducers (T)

• the input of the relays.

• convert the power line signals from a higher level (kiloampere and kilovolt) to a lower level (tens of amperes and volts).

(2) Relays (R)

• process input signals.

• make the decision that a fault has occurred on the transmission line.

(3) Circuit breakers (CB, or B)

• the final link in the fault removal process.

The relay protection device is divided into six parts: sampling unit, processing unit, control unit, operating power supply, identification unit, and execution unit.

## NEW WORDS AND EXPRESSIONS

| | | | |
|---|---|---|---|
| 1. | trip | n. | 跳闸，解扣，松开（弹簧） |
| 2. | block | n. | 封锁 |
| 3. | electromagnet | n. | 电磁铁，电磁体 |
| 4. | parallel | a. | 平行的，并行的 |
| 5. | series | n. | 串联，系列 |
| 6. | amplification | n. | 放大，膨胀 |
| 7. | fault | n. | 故障 |
| 8. | isolate | v. | 隔离，使脱离 |
| 9. | failure | n. | 故障 |
| 10. | remove | v. | 切除，移除 |
| 11. | transistor | n. | 晶体管 |

## PHRASES

**1. in conflict with: 与…冲突**

Example: In general, the reliability requirements of a relay are in conflict with its speed

# Chapter 11　Power System Relay Protection

requirement, sensitivity and reliability are a bunch of contradictions, and a compromise must be made in designing the protection system so as to obtain a reasonable measure of these attributes.

通常继电器的可靠性要求与速动性要求相冲突，灵敏性和可靠性是一对矛盾，在设计保护系统时必须折中考虑，以实现这些属性的合理搭配。

**\*2. be prone to: 有…的倾向，易于**

Example: Earlier models of these relays were prone to frequent component failure under the harsh operating environment.

这些继电器的早期模型在恶劣的运行环境中易于频发元件故障。

**\*3. provided that: 假如，设若**

Example: In distribution systems, the economic aspect almost overrides the technical one owing to the large number of feeders, transformers, etc. provided that basic safety requirements are met.

在配电系统中，由于馈电线、变压器等数目庞大，假如满足了基本的安全要求，则经济方面（的考虑）几乎超过了技术方面。

**\*4. in phase: 同相**

Example: With an internal fault, the current entering end A will be in phase with the current entering end B.

内部故障时，流入 A 端的电流将与流入 B 端的电流同相。

## COMPLICATED SENTENCES

**1. Relays are the logic elements of a protection system that sense the fault and cause the circuit-breaker trip circuits to be energized and the breakers to open their contacts.**

译文：继电器是保护系统的逻辑元件，它感知故障，并使断路器解扣电路带电，断路器打开触点。

说明：由 that 引导的定语从句修饰的主体为 the logic elements 而不是 a protection system。定语从句有两个并列的谓语动词 sense 和 cause，第一个 and 就是连接这两个动词短语的。而 cause 又带有两个并列的宾补结构，中间也是用 and 连接。

**2. A relay used for protection should make its decision as quickly as possible, consistent with other requirements placed upon it.**

译文：在与其他外加要求相容的条件下，用于保护的继电器应该尽可能快的做出决定。

说明：词组 as quickly as possible 表示"尽可能的快"。consistent with other requirements placed upon it 是条件状语从句省略为短语，可补全为：under the condition that it is consistent with other requirements placed upon it.

· 131 ·

3. **The relay should operate consistently for all the faults for which it is designed to operate and should refrain from operating for any other system condition.**

译文：继电器应该始终在按其设计应该动作的所有故障下动作，并且对于任何其他的系统条件避免动作。

说明：定语从句 for which it is designed to operate 修饰的主体是 faults，其句式和含义与 that 引导的宾语从句基本相同，只是在作定语从句时，将 operate 后面的 for 提到了引导词 which 之前。为了便于理解，这个句子可以拆解为：

The relay should operate consistently for all those faults; the relay is designed to operate for those faults, the relay should refrain from operating for any other system condition.

4. **A completely separate arrangement that operates to remove the faulty part should be the main protection that fails to operate.**

译文：这种完全独立的安排意味着一旦主保护未能动作，由它（后备保护）动作去切除故障部分。

说明：should be 部分是虚拟语气的用法，表示假设的情况。

5. **The job of each of these subsystems can best be understood by describing the events that take place from the time of occurrence of a fault to its eventual elimination from the power system.**

译文：每一个子系统的任务，可以通过故障发生时刻到它最终被从电力系统中清除期间所发生的事件来最好地理解。

说明：句中出现了四个 of 和两个 from，这么多的介词连续出现，在翻译时要格外注意句子的通顺。在 that 引导的定语从句中，from the time ... to... 作为 take place 的状语，说明"发生"的时间范围。

\*6. **Trouble may arise due to the magnetizing current inrush on switching operating the relay, and often restraining coils sensitive to third harmonic components of the current are incorporated in the relays.**

译文：由于励磁电流涌入，在操作继电器的开关问题上会出现麻烦，往往把对电流的三次谐波分量敏感的扼流线圈集成到继电器中。

说明：due to 引导的状语短语，在翻译时应提前。on switching 是 trouble 的定语，而 switching 本身带有一个分词短语形式的定语 operating the relay。sensitive to third harmonic components of the current 是 restraining coils 的定语。为了便于理解，这个句子可以拆解为：

Due to the magnetizing inrush current, trouble may arise; trouble is on switching; the switching operates the relay; and often restraining coils are incorporated in the relays; the restraining coils are sensitive to third harmonic components of the current.

## ABBREVIATIONS (ABBR.)

| | | | |
|---|---|---|---|
| 1. | T | transducer | 互感器 |
| 2. | R | relay | 继电器 |
| 3. | CB | circuit breaker | 断路器 |

# Chapter 11  Power System Relay Protection

## SUMMARY OF GLOSSARY

1. 继电器结构　relay structure
   - magnetic furnace　　　　　　　　　　　　　磁炉
   - movable armature　　　　　　　　　　　　　可动衔铁
   - air gap　　　　　　　　　　　　　　　　　　气隙
   - magnetic flux　　　　　　　　　　　　　　　磁通
2. 机电继电器　electromechanical relays
   - plunger type relay　　　　　　　　　　　　活棒式继电器
   - balance-beam type relay　　　　　　　　　平衡木式继电器
   - rotating cup relay　　　　　　　　　　　　转盘式继电器
   - disc relay　　　　　　　　　　　　　　　　圆盘式继电器
3. 继电器种类　relay classes
   - magnitude relay　　　　　　　　　　　　　幅度继电器
   - directional relay　　　　　　　　　　　　定向继电器，极化继电器
   - ratio relay　　　　　　　　　　　　　　　比率继电器
   - differential relay　　　　　　　　　　　差动继电器
   - pilot relay　　　　　　　　　　　　　　　控制 [引示，辅助] 继电器
4. 保护与后备　protection types
   - main protection　　　　　　　　　　　　　主保护
   - primary protection　　　　　　　　　　　主保护
   - backup protection　　　　　　　　　　　　后备保护
   - breaker-failure protection　　　　　　　断路器失效后备保护
   - remote backup protection　　　　　　　　远程后备保护
   - transmission line protection　　　　　　输电线路保护
   - electrical equipment protection　　　　电气设备保护
   - overcurrent protection　　　　　　　　　过电流保护
   - low voltage protection　　　　　　　　　低电压保护
   - overvoltage protection　　　　　　　　　过电压保护
   - power direction protection　　　　　　　功率方向保护
   - impedance distance protection　　　　　阻抗距离保护
   - inter-phase short circuit protection　相间短路保护
   - earth fault protection　　　　　　　　　接地故障保护
   - non-phase operation protection　　　　　非全相运行保护
   - out-of-step protection　　　　　　　　　失步保护
   - loss of excitation protection　　　　　失磁保护
   - differential protection　　　　　　　　　差动保护
   - longitudinal protection　　　　　　　　　纵联差动保护
   - pilot protection　　　　　　　　　　　　纵联保护
   - single-terminal protection　　　　　　　单端量保护
   - distance protection　　　　　　　　　　距离保护
   - current protection　　　　　　　　　　　电流保护
   - ground fault protection　　　　　　　　接地保护
   - electricity protection　　　　　　　　　电（气）量保护
   - non electricity protection　　　　　　　非电量保护
5. 保护技术　protection technology
   - microcomputer protection　　　　　　　　微机保护

wide area protection　　　　　　　　　　　　　广域保护
flexible DC system protection　　　　　　　　　柔性直流系统保护

# EXERCISES

**1. Translate the following words or expressions into Chinese.**

(1) fault　　　　　　(2) block　　　　　　(3) electromagnet
(4) amplification　　(5) contact　　　　　(6) transistor

**2. Translate the following words or expressions into English.**

(1) 继电器　　　　　(2) 灵敏性　　　　　(3) 纵联保护
(4) 后备保护　　　　(5) 保护区　　　　　(6) 差动保护

**3. Fill in the blanks with proper words or expressions.**

(1) In general, the reliability requirements of a relay are in conflict with its _____ requirement, and a _____ must be made in designing the protection system so as to obtain a reasonable measure of these attributes.

(2) The attribute of _____ or _____ means that the relay should operate consistently for all the faults for which it is designed to operate and should refrain from operating for any other system condition.

(3) The concept of _____ helps define the reliability requirements for different protection systems. Each _____ scheme protects a defined area or zone of the power system.

# Word-Building (11)　pre-　预，事先；前

**1. pre-　前缀，表示：预，预先，事先**

| | | | |
|---|---|---|---|
| diction | prediction | n. | 预言，预报 |
| caution | precaution | n. | 预防，警惕，防范 |
| condition | precondition | n. | 前提，先决条件 |
| | | v. | 预处理 |
| arrange | prearrange | v. | 预先安排 |
| design | predesign | v. | 初步[草图]设计，预定[谋] |
| heat | preheat | v. | 预先加热 |
| set | preset | v. | 事先调整，预先设定 |
| determined | predetermined | adj. | 预先规定的 |

**2. pre-　前缀，表示：前，在……之前的**

| | | | |
|---|---|---|---|
| fix | prefix | n. | 前缀 |
| disturbance | pre-disturbance | n. | 扰动前 |

# Chapter 12
# Insulation and Grounding

## Part 1　Introduction Electrical Insulation

　　The insulation of an electric power system is of critical importance from the standpoint of service continuity. Probably more major equipment troubles are traceable to insulation failure than to any other cause.

　　Insulation is required to keep electrical conductors separated from each other and other nearby objects. Ideally, insulation should be nonconducting, for then currents are restricted to the intended conductors. However, insulation does conduct some current and so must be regarded as a material of very high resistivity. * In many applications, the current flow due to conduction through the insulation is so small that it may be entirely neglected. In some instances, the conduction currents, measured by very sensitive instruments, serve as a test to determine the suitability of the insulation for use in service.

　　The choice of material is dictated by the requirements of the particular application and by cost.

　　In residences, the conductors used in branch circuits and the cords to appliances may be insulated with rubber or plastics of several different kinds. Such materials can withstand necessary bending and have relatively low electric stress.

　　High-voltage cables are subjected to extreme voltage stress; in some cases, several hundred kilovolts are impressed across a few centimeters of insulation. They must be manufactured in large sections and must be sufficiently flexible to permit pulling into ducts of small cross sections. The insulation may be oil-impregnated paper, varnished cambric, or synthetic materials such as polyethylene.

　　The coils of generators and motors may be insulated with various tapes. Some of these are made of thin sheets of mica held together by a binder, and others are of fiber glasses impregnated with insulating varnish. The insulation must be able to withstand quite high operating temperatures, extreme mechanical forces, and vibration.

　　The insulation on power-transformer windings is commonly paper tape and pressboard operated under oil. The transformer insulation is subjected to high electric stress and large mechanical forces. * The shape and arrangement of conducting metal parts are of particular concern in transformer design.

　　Overhead lines are supported on porcelain insulators. Between the supports air serves as insulation. Porcelain is chosen because of its resistance to deterioration when exposed to the weather, its high dielectric strength, and its ability to wash clean in rain.*

Although insulating materials are very stable under ordinary circumstances, they may change radically in characteristics under extreme conditions of voltage stress or temperature or under the action of certain chemicals.* Such changes may, in local regions, result in the insulating material becoming highly conductive. Unwanted current flow brings about intense heating and the rapid destruction of the insulating material. These insulation failures account for a high percentage of the equipment troubles in electric power systems.

The cost of equipment and the system as a whole is closely related to the insulation level that has been adopted for the network. The choice of insulation level should take into account both the probability of occurrence and the severity of stress that the system may be subjected to. Therefore, the performance of any protective device inserted into the system should be defined as quantitatively as possible.

## Part 2　Insulation Coordination

Insulation coordination is the correlation of insulation of electrical equipment with the characteristics of protective devices such that the insulation is protected from excessive over-voltages.* In a substation, for example, the insulation of transformers, circuit breakers, bus supports, etc., should have insulation strength in excess of the voltage provided by protective devices.

Electric systems' insulation designers have two options available to them: ① choose insulation levels for components that would withstand over-voltages, ② devise protective devices that could limit over-voltages there. The first alternative is unacceptable especially for extra-high voltage (EHV) and ultra high voltage (UHV) operating levels because of the excessive insulation required. Hence, there has been a great incentive to develop and use protective devices. The actual relationship between the insulation level and protective levels is a question of economics. Conventional methods of insulation coordination provide a margin of protection between electrical stress and electrical strength based on predicated maximum overvoltage and minimum strength, the maximum strength being allowed by the protective devices. "Insulation level" is defined by values of the test voltages that the insulation of equipment under test must be able to withstand.

In the earlier days of electric power, insulation levels commonly used were established on the basis of experience gained by utilities. As laboratory techniques improved, so that different laboratories were in closer agreement on test results, an international joint committee, the Nema-Nela Committee on Insulation Coordination, was formed and was charged with the task of establishing insulation strength of all classes of equipment and establishing levels for various voltage classifications.* In 1941 a detailed document was published giving basic insulation levels for all equipment in operation at that time. The presented tests included standard

# Chapter 12   Insulation and Grounding

impulse voltages and 1-minute power frequency tests.

In today's systems for voltages up to 245kV, the tests are still limited to lightning impulses and 1-minute power frequency tests. Above 300kV, tests include in addition to lightning impulse and the 1-minute power frequency tests, the use of switching impulse voltages.* Impulse test voltages (1.2/50µs) are usually referred to as the "Basic Insulation Levels" (BILs), defined as the impulse voltage which the insulation of any electrical equipment for a given rated voltage must be able to withstand, also commonly known as "Impulse Withstand Level".*

**Statistical Approach to Insulation Coordination**

In the early days insulation levels for lightning surges were determined by evaluating the 50% flashover values for all insulations and providing sufficiently high withstand levels for all insulations. This approach is difficult to apply at EHV and UHV levels, particularly for external insulations.

Present-day practices of insulation coordination rely on a statistical approach that relates directly the electrical strength. This approach requires a knowledge of the distribution of both the anticipated stresses and the electrical strength.

The statistical nature of over-voltages, in particular switching over-voltages, makes it necessary to compute a large number of over-voltages in order to determine with some degree of confidence the statistical over-voltages on a system.* The EHV and UHV systems employ a number of nonlinear elements, but with today's availability of digital computers, the distribution of over-voltages can be calculated.

The dielectric strengths of external self-restoring insulations are determined through tests carried out in laboratories and the data that the development of electrical breakdown is governed by statistical laws and has a random character; furthermore it is found that the distribution of breakdowns for a given gap follows with some exceptions approximately normal or Gaussian distribution, as does the distribution of over-voltages on the system.* For the purpose of coordinating the electrical stresses with electrical strengths it is convenient to represent the overvoltage distribution in the form of the probability density function and the insulation breakdown probability by the cumulative distribution function. The knowledge of these distributions enables us to determine the "risk of failure".

In engineering practice it would become uneconomical to use the complete distribution functions for the occurrence of overvoltage and the withstanding of insulation and a compromise solution is accepted for guidance.

**Correlation between Insulation and Protection Levels**

"The protection level" is established in a similar manner to the "insulation level". The basic difference is that the insulation of protective devices must not withstand the applied voltage. The concept of correlation between insulation and protection levels can be readily understood by considering a simple example of an insulator string being protected by a spark gap, the gap (of lower breakdown strength) protecting the insulator string.

Both gaps are subjected to the same overvoltage. In practice, the protective gap will break down before the insulation and will cause a reduction (to a safe limit) in overvoltage reaching the protected insulation.

# Part 3　Grounding

Grounding is defined as a conducting connection, whether intentional or accidental, by which an electric circuit or equipment is connected to the earth, or to some conducting body of a relatively large extent that serves in place of the earth.* It is used for establishing and maintaining the potential of the earth (or of the conducting body) or approximately that potential, on conductors connected to it; and for conducting ground current to and from the earth (or the conducting body).*

The most important reason for grounding is safety. Two important aspects of grounding requirements with respect to safety and one with respect to power quality are

**1. Personnel safety**

Personnel safety is the primary reason that all equipment must have a safety equipment ground. This is designed to prevent the possibility of high touch voltages when there is a fault in a piece of equipment. The touch voltage is the voltage between any two conducting surfaces that can be simultaneously touched by an individual. The earth may be one of these surfaces.

There should be no "floating" panels or enclosures in the vicinity of electric circuits. In the event of insulation failure or inadvertent application of moisture, any electric charge that appears on a panel, enclosure, or raceway must be drained to "ground" or to an object that is reliably grounded.

**2. Grounding to assure protective device operation**

A ground fault return path to the point where the power source neutral conductor is grounded is an essential safety feature.

An insulation failure or other fault that allows a phase wire to make contact with an enclosure will find a low-impedance path back to the power source neutral. The resulting overcurrent will cause the circuit breaker or fuse to disconnect the faulted circuit promptly.

An effective grounding path (the path to ground from circuits, equipment, and conductor enclosures) shall

1) Be permanent and continuous;

2) Have the capacity to conduct safely any fault current likely to be imposed on it;

3) Have sufficiently low impedance to limit the voltage to ground and to facilitate the operation of the circuit protective devices in the circuit;

# Chapter 12  Insulation and Grounding

4) Not have the earth as the sole equipment ground conductor;

**3. Noise control**

Noise control includes transients from all sources. This is where grounding relates to power quality. Grounding for safety reasons defines the minimum requirements for a grounding system. Anything that is done to the grounding system to improve the noise performance must be done in addition to the minimum requirements.

The primary objective of grounding for noise control is to create an equipotential ground system. Potential differences between different ground locations can stress insulation, create circulating ground currents in low-voltage cables, and interfere with sensitive equipment that may be grounded in multiple locations.

Ground voltage equalization of voltage differences between parts of an automated data processing (ADP) grounding system is accomplished in part when the equipment grounding conductors are connected to the grounding point of a single power source. However, if the equipment grounding conductors are long, it is difficult to achieve a constant potential throughout the grounding system, particularly for high-frequency noise. Supplemental conductors, ground grids, low-inductance ground plates, etc., may be needed for improving the power quality. These must be used in addition to the equipment ground conductors, which are required for safety, and not as a replacement for them.*

## NEW WORDS AND EXPRESSIONS

| | | | |
|---|---|---|---|
| 1. | resistivity | *n.* | 电阻系数，电阻率 |
| 2. | dictate | *vt.* | 指导，支配 |
| 3. | cord | *n.* | 软线，塞绳，电线电缆 |
| 4. | withstand | *vt.* | 抵挡，经受住 |
| 5. | bending | *n.* | 弯曲（度） |
| 6. | electric stress | | 静电应力，电介质应力 |
| 7. | duct | *n.* | 电线、电缆的管道 |
| 8. | oil-impregnated | *adj.* | 浸油的，注入油的 |
| 9. | varnished cambric | | 细漆布，涂漆细麻布 |
| 10. | synthetic material | | 合成材料，人造材料 |
| 11. | polyethylene | *n.* | 聚乙烯 |
| 12. | mica | *n.* | 云母 |
| 13. | binder | *n.* | 黏合剂，胶 |
| 14. | fiber glass | *n.* | 玻璃纤维（耐热绝缘材料） |
| 15. | insulating varnish | | 绝缘清漆 |
| 16. | vibration | *n.* | 振动，颤动 |
| 17. | pressboard | *n.* | 纸板 |
| 18. | porcelain insulator | | 瓷绝缘子 |
| 19. | deterioration | *n.* | 老化，退化，变坏 |
| 20. | dielectric strength | | 电介质强度，绝缘强度 |
| 21. | devise | *vt.* | 设计，发明 |

| 22. | power frequency | | 工频 |
| 23. | statistical approach | | 统计法 |
| 24. | surge | n. | 电流急冲，电涌 |
| 25. | flashover | n. | 闪络，跳火 |
| 26. | statistical nature | | 统计特性 |
| 27. | nonlinear | adj. | 非线性的 |
| 28. | breakdown | n. | 击穿 |
| 29. | normal distribution | | 正态分布 |
| 30. | Gaussian distribution | | 高斯分布 |
| 31. | probability density | | 概率密度，几率密度 |
| 32. | cumulative distribution | | 累积分布 |
| 33. | compromise solution | | 折中方法 |
| 34. | arrester | n. | 避雷器 |
| 35. | spark gap | | 火花隙 |
| 36. | personnel safety | | 人身安全 |
| 37. | inadvertent | adj. | 不注意的，疏忽的，无意的 |
| 38. | facilitate | vt. | 推动，帮助，促进 |
| 39. | equipotential | adj. | 等电位的 |

## PHRASES

**1. be subjected to: 遭受，承受**

Example: High-voltage cables are subjected to extreme voltage stress.
高压电缆承受极大的电压应力。

**2. in excess of: 超过，较…为多**

Example: In a substation, for example, the insulation of transformers, circuit breakers, bus supports, etc., should have insulation strength in excess of the voltage provided by protective devices.
例如，在变电站中，变压器、断路器、母线支柱等的绝缘应该具有比保护装置提供的电压高的绝缘强度。

**3. on the basis of: 以…为基础**

Example: In the earlier days of electric power, insulation levels commonly used were established on the basis of experience gained by utilities.
在电力发展的早期，通常采用的绝缘等级是以电力公司获取的经验为基础建立的。

**4. a knowledge of: 对…的总体了解**

Example: This approach requires a knowledge of the distribution of both the anticipated stresses and the electrical strength.
该方法需要对预期应力和电场强度的分布都有总体了解。

# Chapter 12　Insulation and Grounding

**5. for the purpose of: 为了，因…起见**

Example: For the purpose of coordinating the electrical stresses with electrical strengths it is convenient to represent the overvoltage distribution in the form of the probability density function and the insulation breakdown probability by the cumulative distribution function.

为了使静电应力和电场强度配合，最好用概率密度函数的形式表示过电压，用累积分布函数表示绝缘击穿概率。

**6. in a similar manner to: 以与…相同的方式**

Example: "The protection level" is established in a similar manner to the "insulation level".
"保护水平"以与"绝缘水平"相同的方式建立。

**7. with respect to: 关于，至于**

Example: Two important aspects of grounding requirements with respect to safety and one with respect to power quality are...

接地要求关于安全的两个重要方面和关于电能质量的一个重要方面是……

**8. in the vicinity of: 在…附近，邻近**

Example: There should be no "floating" panels or enclosures in the vicinity of electric circuits.

电路附近不应有（电位）"悬浮"的面板或围栏。

**9. interfere with: 干涉，干扰**

Example: Potential differences between different ground locations can interfere with sensitive equipment that may be grounded in multiple locations.

不同接地点之间的电位差会干扰多处同时接地的敏感设备。

## COMPLICATED SENTENCES

**1. However, insulation does conduct some current and so must be regarded as a material of very high resistivity.**

译文：然而，绝缘部分确实传导一些电流，因此必须被认为是一种具有很高电阻率的材料。

说明：句中 does 表示强调，可译为"确实"。and 连接的两个谓语动词 does 和 must 共属一个主语 insulation。so 在此作为副词使用。

**2. The transformer insulation is subjected to high electric stress and large mechanical forces.**

译文：变压器的绝缘承受很高的静电应力和很大的机械压力。

说明：谓语表意动词 subject 带有两个并列的介宾短语，也可以看作两个并列动词短语的省略形式，补全应为：The transformer insulation is subjected to high electric stress and is subjected to large mechanical forces.

**3. Porcelain is chosen because of its resistance to deterioration when exposed to the weather, its high dielectric strength, and its ability to wash clean in rain.**

译文：选择陶瓷是因为它暴露在气象条件中时的抗老化能力、绝缘强度大以及雨中的清洗保洁能力。

说明：because of 带有三个并列的以 its 开头的宾语，在翻译时为了避免重复，这三个 its 不必都翻译。when 引导的时间状语从句修饰的主体是 its resistance to deterioration。另外，按照中文表述习惯，形如 its high dielectric strength 的短语常常译为主谓结构，即按 its dielectric strength is high 来翻译。

**4. Although insulating materials are very stable under ordinary circumstances, they may change radically in characteristics under extreme conditions of voltage stress or temperature or under the action of certain chemicals.**

译文：虽然在平常情况下绝缘材料很稳定，但在极端的电压应力或温度条件下，或者在某些化学作用下，其特性会完全改变。

说明：they may change radically in characteristics 部分相当于 their characteristics may change radically，而且按照后者翻译，往往更符合中文表述习惯。

**5. Insulation coordination is the correlation of insulation of electrical equipment with the characteristics of protective devices such that the insulation is protected from excessive over-voltages.**

译文：绝缘配合是电气设备绝缘与保护装置特性的相互关系，为的是保护绝缘免受过度的过电压。

说明：the correlation of...with... 表示"…和…的相互关系"。such that 在这里引导一个表示目的的从句。词组 be protected from 表示"保护，使免受"。

**6. As laboratory techniques improved, so that different laboratories were in closer agreement on test results, an international joint committee, the Nema-Nela Committee on Insulation Coordination, was formed and was charged with the task of establishing insulation strength of all classes of equipment and establishing levels for various voltage classifications.**

译文：随着实验室技术的改进，因而不同实验室的结果近似一致，关于绝缘配合的 Nema-Nela 委员会成立并负责为所有种类设备确定绝缘强度，确定各种电压分级水平。

说明：different laboratories were in closer agreement on test results 相当于 test results of different laboratories were in closer agreement，而且按照后者的语序进行翻译，似乎更符合中文表述习惯。其中 in closer agreement on 表示"在…方面近似一致"。be charged with the task of 相当于 be charged in，表示"承担…任务，负责…"。其中 task 带有两个修饰短语，分别是动名词形式 establishing... 和动词不定式形式 to establish。

**7. Above 300kV, tests include in addition to lightning impulse and the 1-minute power frequency tests, the use of switching impulse voltages.**

译文：300kV 以上，除了雷电脉冲和 1min 工频试验之外，试验还包括开关脉冲电压的使用。

# Chapter 12　Insulation and Grounding

说明：in addition to 表示"除了……之外"，引导一个状语短语，在翻译时应提前。为了便于理解，这个句子可以拆解为：

Above 300kV, in addition to lightning impulse and the 1-minute power frequency tests, tests include the use of switching impulse voltages.

**8. Impulse test voltages are usually referred to as the "Basic Insulation Levels", defined as the impulse voltage which the insulation of any electrical equipment for a given rated voltage must be able to withstand, also commonly known as "Impulse Withstand Level".**

译文：脉冲试验电压通常被称为"基本绝缘水平"，定义为给定电压等级下的任何电气设备都必能承受的脉冲电压，也通常称为"脉冲耐受水平"。

说明：be referred to as 表示"作为…被提及"，be known as 表示"作为…被知晓"，为了翻译通顺，二者均可译为"被称为"。

**9. The statistical nature of over-voltages, in particular switching over-voltages, make it necessary to compute a large number of over-voltages in order to determine with some degree of confidence the statistical over-voltages on a system.**

译文：过电压（尤其是开关过电压）的统计特性，有必要计算大量的过电压，以有一定信心确定系统的统计过电压。

说明：in particular 表示"尤其，特别"，make it necessary 表示"使…有必要"，为了翻译通顺，可简译为"有必要…"。其中 it 代指 necessary 后面的内容，以同位语的形式出现。with some degree of confidence 表示"带有某种程度的信心"，是状语短语，翻译时可以从所修饰的动词后面移走，一般应提前。为了便于理解，该句子可以拆解为：

The statistical nature of over-voltages (in particular switching over-voltages) make it necessary; in order to determine the statistical over-voltages on a system with some degree of confidence, it is necessary to compute a large number of over-voltages.

**10. The dielectric strengths of external self-restoring insulations are determined through tests carried out in laboratories and the data that the development of electrical breakdown is governed by statistical laws and has a random character; furthermore it is found that the distribution of breakdowns for a given gap follows with some exceptions approximately normal or Gaussian distribution, as does the distribution of over-voltages on the system.**

译文：外部自恢复绝缘体的电介质强度通过实验室进行的试验来确定，电气击穿的数据按统计规律控制，具有随机性。此外，还发现对于给定的气隙而言，击穿的分布除了个别例外近似遵循正态分布或高斯分布，系统中的过电压分布也是这样。

说明：短语 that the development of electrical breakdown 的作用相当于一个同位语从句，与 the data 所指相同。with some exceptions 是状语短语，意思是"具有某些例外"，在翻译时可以从动词后移走，一般应提前。以 as does 为引导的部分为一个倒装句。

**11. Grounding is defined as a conducting connection, whether intentional or accidental, by which an electric circuit or equipment is connected to the earth, or to some conducting body of a relatively large extent that serves in place of the earth.**

译文：接地定义为电路或电气设备连接到大地或适合代替大地的较大导体的故意的或意外的导通连接。

说明：介词＋连词构成的引导词 by which 引导的定语从句和短语 whether intentional or accidental 修饰的主体都是 a conducting connection。句中 to some conducting body 与 is connected to the earth 并列，前面省略了动词 is connected 部分。

**12. It is used for establishing and maintaining the potential of the earth or approximately that potential, on conductors connected to it; and for conducting ground current to and from the earth.**

译文：它用于在与之相连的导体上建立和维持大地的电位或近似该电位，还用于与大地之间交换接地电流。

说明：on conductors connected to it 是状语短语，翻译时可提前。to and from 表示有去有回，可以理解为双向交换。

**13. These must be used in addition to the equipment ground conductors, which are required for safety, and not as a replacement for them.**

译文：在安全所要求的设备接地导体之外，这些必须用上，不是作为其替代品。

## ABBREVIATIONS (ABBR.)

| | | | |
|---|---|---|---|
| 1. | EHV | extra high voltage | 超高压 |
| 2. | UHV | ultra high voltage | 特高压 |
| 3. | BIL | basic insulation levels | 基本绝缘水平 |
| 4. | ADP | automated data processing | 自动数据处理 |

## SUMMARY OF GLOSSARY

1. 应力与强度　stress and strength
   electric stress　　　　　　　　　　静电应力，电介质应力
   voltage stress　　　　　　　　　　电压应力
   mechanical force　　　　　　　　　机械压力
   electrical strength　　　　　　　　电场强度
   dielectric strength　　　　　　　　绝缘强度，电介质强度
   insulation strength　　　　　　　　绝缘强度
2. 统计规律　statistical laws
   random distribution　　　　　　　　随机分布
   normal distribution　　　　　　　　正态分布
   Gaussian distribution　　　　　　　高斯分布

# Chapter 12  Insulation and Grounding

3.　统计函数　statistical functions
　　probability density function　　　　　　概率密度函数
　　cumulative distribution function　　　　累积分布函数

# EXERCISES

**1. Translate the following words or expressions into Chinese.**

(1) appliance　　　　　(2) surge　　　　　　(3) electric stress
(4) flashover　　　　　(5) statistical nature　(6) arrester
(7) stator field　　　　(8) enclosure　　　　(9) equipotential

**2. Translate the following words or expressions into English.**

(1) 电介质强度　　　　(2) 裕度　　　　　　(3) 瓷绝缘子
(4) 闪电　　　　　　　(5) 工频　　　　　　(6) 击穿
(7) 人身安全　　　　　(8) 火花隙　　　　　(9) 电缆管道

**3. Fill in the blanks with proper words or expressions.**

(1) The transformer insulation is subjected to high electric ＿＿＿＿ and large mechanical ＿＿＿＿.

(2) There are two important aspects of grounding requirements with respect to ＿＿＿＿ and one with respect to ＿＿＿＿.

(3) For the purpose of coordinating the electrical ＿＿＿＿ with electrical ＿＿＿＿ it is convenient to represent the ＿＿＿＿ distribution in the form of the probability density function and the insulation ＿＿＿＿ probability by the cumulative distribution function.

(4) "The protection level" is established in a similar manner to the "＿＿＿＿ level".

# Word-Building (12)　co-; inter-　互，共

**1. 前缀 co- 表示：共，同，相互，联合，伴同**

| | | | |
|---|---|---|---|
| exist | coexist | vi. | 共存 |
| operate | cooperate | vi. | 合作，协作 |
| axial | coaxial | adj. | 同轴的，共轴的 |
| efficient | coefficient | adj. n. | 共同作用的；系数，折算率，共同作用 |
| ordination | coordination | n. | 配合；协调 |
| relation | correlation | n. | 相互关系，相关（性） |
| author | co-author | n. | 合著者 |

| star | co-star | n. | 联合主演 |
| worker | co-worker | n. | 合作者，同事，帮手 |

## 2. 前缀 inter- 表示：交互，互相

| action | interaction | n. | 交互作用，交感 |
| change | interchange | v. | 相互交换 |
| connect | interconnect | vt. | （使）互相连接 |
| net | internet | n. | 互联网络 |
| tripping | intertripping | n. | 联锁跳闸 |

# Chapter 13
# Overvoltage and Lightning Shielding

## Part 1  Overvoltage and Its Classification

Disturbances of electric power transmission and distribution systems are frequently caused by some kinds of transient voltages whose amplitudes may greatly exceed the peak values of normal AC of operating voltage.

Concretely, power systems are often subjected to over-voltages that have their origin in atmospheric discharges in which case they are called external overvoltages, or they are generated internally by connecting or disconnecting the system, or due to the system fault initiation or extinction.* The latter type is called *internal overvoltages*.

The magnitude of the external over-voltages, e.g. lightning over-voltages, remains essentially independent of the system's design, whereas internal over-voltages, e.g. switching over-voltages, increase with increasing the operating voltage of the system. Hence, with increasing the system's operating voltage a point is reached when the switching over-voltages became the dominant factor in designing the system's insulation up to approximately 300kV.* The system's insulation has to be designed to withstand primarily lightning surges. Above that voltage, both lightning and switching surges have to be considered. For UHV systems, 765kV and above switching over-voltages in combination with insulator contamination becomes the predominant factor in the insulation design.

Over-voltages represent a major threat to security and continuity of supply. Fortunately, a controlled limitation of overvoltage surges is possible by protective devices. For the economic design of equipment and safe operation of power systems, detailed knowledge of types and sources of over-voltages on power systems is required.

The over-voltages can be first divided into external over-voltages and *internal over-voltages* according to the locations of sources as discussed above.

**1. External over-voltages**

External over-voltages are generated by sources that are external to the power system network. Their magnitude is essentially independent of the system. Two types of external overvoltage have been identified.

(1) Lightning over-voltages. Lightning over-voltages are characterized by very high peak currents and relatively low energy content. They are responsible for nearly half of all short

circuits on lines in systems of 300kV and above. Up to approximately 300kV, the system insulation has to be designed primarily to withstand lightning surges. Lightning over-voltages originate from lightning strokes hitting the phase wires of overhead lines or the busbars of outdoor substations. The amplitudes are very high, usually in the order of 1000kV and even more into the transmission line; each stroke is then the maximum insulation strength of the overhead line.* The rate of voltage rise of such a traveling wave is at its origin directly proportional to the steepness of the lightning current, which may exceed 100kA/μs, as the voltage levels may simply be calculated by the current multiplied by the effective surge impedance of the line.* Too high voltage levels are immediately chopped by the breakdown of the insulation and therefore traveling waves with steep wave fronts and even steeper wave tails may stress the insulation of power transformers or other HV equipment severely. Lightning protection systems, surge arresters, and the different kinds of losses will dampen and distort the traveling waves, and therefore lightning over-voltages with very different waveshapes are present within the transmission system.

(2) Nuclear and non-nuclear electromagnetic pulses (NEMP, NNEMP). NEMP is also an external source of power system over-voltages. They are characterized by very short front times (<10nm) and tens of kiloamperes in amplitude. Their occurrence is fortunately rare and protecting against them is usually not considered in power systems design. NNEMP has been reported to affect power systems during magnetic storms associated with exceptional solar disturbances.

**2. Internal over-voltages**

Unlike external over-voltages, internal over-voltages are principally determined by the system configuration and its parameters. They are mainly due to switching operations and following fault conditions. According to their duration, which ranges from a few hundred microseconds to several seconds, three main categories of internal over-voltages can be defined: ① *temporary over-voltages*, if they are oscillatory of power frequency or harmonics, ② *switching over-voltages*, if they are heavily damped and of short duration, and ③ *steady state over-voltages*, if they are of long duration. Temporary over-voltages occur almost without exception under no load or very light load conditions. Because of their common origin, the temporary and switching over-voltages occur together and their combined effect has to be taken into account in the design of HV systems insulation.

(1) Transient switching surges. Compared with lightning surges, switching surges contain a much higher amount of energy but are of lower amplitude. Switching over-voltages are caused by switching phenomena. Their amplitudes are always related to the operating voltage and the shape is influenced by the impedances of the system as well as by the switching conditions. The rate of voltage rise is usually slower, but it is known today that the waveshapes can also be very dangerous to different insulation systems. These types of over-voltages are also effective in the LV distribution systems, where they are either produced by the usually

# Chapter 13  Overvoltage and Lightning Shielding

current-limiting switches or where they have been transmitted from the HV distribution systems.* Here they may often cause a breakdown of electronic equipment, as they can reach amplitudes of several kilovolts, and it should be mentioned that the testing of certain LV apparatus with transient voltages is a need today.

(2) Temporary over-voltages. A definition adopted by many authors describes a temporary overvoltage (TOV) as an oscillatory phase-to-ground or phase-to-phase overvoltage of relatively long duration at a given location which is undamped or weakly damped in contrast to switching and lightning over-voltages which are usually highly or very highly damped and of short times.* Temporary over-voltages have also been classified into three groups according to whether the frequency of oscillation is lower, equal to or higher than the working voltage frequency.

(3) Steady-state over-voltages. Steady-state over-voltages are of supply frequency and are sustained for long periods. They are often related to earthing and neutral arrangements within the system. Examples are the neutral displacement in badly designed star-connected voltage transformers, and arcing ground phenomena in systems with insulated neutrals and the resonance phenomena appearing as a result of the open-circuiting of one or two phases in a three-phase system which can be initiated by faulty circuit breakers or broken conductors.* Most systems are designed so that such overvoltages do not occur.

## Part 2  Lightning and Its Hazards

The condition for generating lightning is the accumulation and formation of polarity in thunderstorm clouds. During stormy weather, a charge separation takes place inside clouds, so that positive charges move to the upper part of the cloud while negative charges stay below (Figure 14.1).

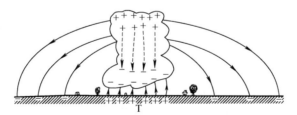

**Figure 14.1**  Electric fields created by a thundercloud

As more and more positive charges move upward within the cloud, the electric field below the cloud becomes more and more intense. Ultimately, it reaches the critical ionization level where the air begins to break down. Ionization takes place first at the tips of church spires and the top of high trees, and may sometimes give rise to a bluish light.

When the electric field becomes sufficiently intense, lightning will suddenly strike from cloud to earth. A single stroke may involve a charge transfer of from 0.2 to 20 coulombs, under a difference of potential of several hundred million volts. The current per stroke rises to a peak in one or two microseconds and falls to half its peak value in about 50μs. What is visually observed as a single stroke is often composed of several strokes following each other in rapid succession. The total discharge time may last as long as 200ms. Discharges also occur between positive and negative charges within the cloud, rather than between the base of the cloud and the ground.

Lightning has a large lightning current amplitude, large lightning steepness, strong impact, and high impact overvoltage. Its characteristics are closely related to its destructive nature.

Lightning has a variety of destructive effects such as electrical, thermal, and mechanical properties, all of which can have extremely serious consequences.

(1) Fire and explosion. High-temperature arc of direct lightning discharge, secondary discharge, huge lightning current, and ball lightning invasion can directly cause fire and explosion; impulse voltage breakdown of electrical equipment insulation damage can indirectly cause fire and explosion.

(2) Electric shock. Cumulus clouds directly to the human body discharge, secondary discharge, ball lightning strike, touch voltage, and step voltage generated by lighting current can directly cause people electric shock; damage to electrical equipment insulation from lightning strikes can also cause electric shocks.

(3) Damage to equipment and facilities. The vaporization force, electrostatic force, and electromagnetic force generated by lightning strikes with high voltage and large currents can destroy important electrical devices, buildings, and other facilities.

(4) Large-scale power outages. The destruction of power equipment or power lines may lead to large-scale blackouts. The lightning trip-out rate is a comprehensive index to measure the lightning protection performance of lines.

## Part 3  Lightning-proof Measures

In order to prevent and reduce lightning damage to buildings, electrical and electronic systems, to protect people's lives and property safety, and to protect electrical and electronic system equipment safety, it is necessary to take some lightning protection measures. There are four main types of lightning-proof measures:

(1) Direct lightning stroke protection. The installation of the lightning rod, lightning shield line, lightning net, and lightning protection belt is the main measure of direct lightning stroke protection-The protection range of the lightning rod is calculated by the Rolling Sphere Method.

(2) Secondary discharge protection. In order to prevent secondary discharge, whether in the air or underground, it is necessary to ensure that there is a sufficient safe distance between the lightning receiver, down lead, grounding device, and the

# Chapter 13  Overvoltage and Lightning Shielding

adjacent conductor.

(3) Inductive lightning protection. In order to prevent the danger of electrostatic induction lightning, the uncharged metal equipment and metal structures in the building should be connected and grounded.

(4) Lightning shock wave protection. In order to prevent the lightning shock wave from invading the power distribution device, a valve arrester can be installed at the line inlet end.

The main types of arresters are tubular arresters, valve arresters, zinc oxide arresters, etc.

The tubular arrester is a protection gap with high arc extinguishing ability. The arc extinguishing ability of the tube-type arrester is related to the size of the power frequency continuous current, which is mostly used for lightning protection on power supply lines.*

The valve arrester is composed of a spark gap and valve resistance. The valve resistance made of silicon carbide can effectively prevent lightning and high voltage and protect the equipment.

Zinc oxide arrester is a kind of lightning protection equipment with superior protection performance, lightweight, pollution resistance, and stable performance. The difference between it and the traditional arrester is that it has no discharge gap and utilizes the nonlinear characteristics of zinc oxide to discharge and break.

## NEW WORDS AND EXPRESSIONS

| 1. | concretely | adv. | 具体地 |
| 2. | extinction | n. | 消灭，消除 |
| 3. | dominant factor | | 主要因素，基本因素 |
| 4. | contamination | n. | 玷污，污染，污染物 |
| 5. | predominating | adj. | 占优势的 |
| 6. | energy content | | 能含量；内能 |
| 7. | lightning stroke | | 雷击 |
| 8. | steepness | n. | 陡度 |
| 9. | chop | vt. | 砍，切碎 |
| 10. | steep wave front | | 陡波前沿 |
| 11. | wave tail | | 波尾，电波信号的尾部 |
| 12. | damp | vt. | 阻尼，使衰减 |
| 13. | magnetic storm | | 磁暴（太阳黑子引起地磁场扰动） |
| 14. | resonance | n. | 谐振，共振 |
| 15. | charge separation | | （等离子）电荷分离 |
| 16. | intense | adj. | 很大的，十分强烈的 |
| 17. | ionization | n. | 离子化，电离 |
| 18. | tip | n. | 顶，尖端 |
| 19. | church spire | | 教堂尖顶 |

| 20. | bluish | adj. | 带蓝色的 |
| 21. | coulomb | n. | 库仑（电量单位） |
| 22. | lightning steepness | | 雷电流陡度 |
| 23. | cumulus clouds | | 积云 |
| 24. | step voltage | | 跨步电压 |
| 25. | vaporization force | | 汽化力 |
| 26. | electrostatic force | | 静电力 |
| 27. | electromagnetic force | | 电磁力 |
| 28. | lightning trip-out rate | | 雷击跳闸率 |
| 29. | comprehensive | adj. | 综合性的，全面的 |
| 30. | lightning rod | | 避雷针 |
| 31. | lightning shield line | | 避雷线 |
| 32. | lightning net | | 避雷网 |
| 33. | lightning protection belt | | 避雷带 |
| 34. | rolling sphere method | | 滚球法 |
| 35. | electrostatic | adj. | 静电的 |
| 36. | inlet | n. | 进口，入口 |
| 37. | tubular arrester | | 管型避雷器 |
| 38. | valve arrester | | 阀型避雷器 |
| 39. | zinc oxide arrester | | 氧化锌避雷器 |
| 40. | spark | n. | 火花，火星 |
| 41. | silicon carbide | | 碳化硅 |
| 42. | utilize | v. | 利用，使用 |
| 43. | nonlinear | adj. | 非线性的 |

## PHRASES

**1. in combination with: 与…结合**

Example: For UHV systems, 765kV and above switching over-voltages <u>in combination with</u> insulator contamination becomes the predominant factor in the insulation design.

对于特高压系统，765kV 及以上的操作过电压与绝缘子污染相结合，成为绝缘设计中的支配性因素。

**2. without exception: 毫无例外地，一律**

Example: Temporary over-voltages occur almost <u>without exception</u> under no load or very light load conditions.

临时过电压几乎毫无例外地发生在无载或非常轻载的条件下。

**3. take into account: 重视，考虑**

Example: Because of their common origin, the temporary and switching over-voltages occur together and their combined effect has to be <u>taken into account</u> in the design of HV sys-

# Chapter 13　Overvoltage and Lightning Shielding

tems insulation.

由于有共同的起因，临时过电压和操作过电压同时发生，其综合影响在高压系统绝缘设计中必须引起重视。

**4. in rapid succession: 紧接着，接连地**

Example: What is visually observed as a single stroke is often composed of several strokes following each other in rapid succession.
看上去是单个雷击，往往由相互之间紧紧跟随的多个雷击组合而成。

**5. rather than: 胜于，而不是**

Example: Discharges also occur between positive and negative charges within the cloud, rather than between the base of the cloud and the ground.
放电也会出现在云层内部的正负电荷之间，多于出现在云层底部和大地之间。

**6. be related to: 与…有关**

Example: Its characteristics are closely related to its destructive nature.
它的特点与其破坏性有紧密的关系。

**7. take measures: 采取措施**

Example: In order to prevent and reduce lightning damage to buildings, electrical and electronic systems, to protect people's lives and property safety, and to protect electrical and electronic system equipment safety, it is necessary to take some lightning protection measures.
为防止和减少雷电对建筑物、电气、电子系统造成的危害，保护人民的生命和财产安全，保护电气、电子系统设备安全，有必要采取一些防雷措施。

**8. whether...or...: 不论是…还是…**

Example: In order to prevent secondary discharge, whether in the air or underground, it is necessary to ensure that there is a sufficient safe distance between the lightning receiver, down lead, grounding device, and the adjacent conductor.
为了防止二次放电，不论是空气中还是地下，都必须保证接闪器、引下线、接地装置与邻近导体之间有足够的安全距离。

**9. prevent from: 阻止，防止**

Example: In order to prevent the lightning shock wave from invading the power distribution device, a valve arrester can be installed at the line inlet end.
为了防止雷电冲击波侵入变配电装置，可在线路引入端安装阀型避雷器。

**10. be composed of: 由…组成**

Example: The value arrester is composed of a spark gap and valve resistance.
阀型避雷器由火花间隙及阀片电阻组成。

# COMPLICATED SENTENCES

**1. Concretely, power systems are often subjected to over-voltages that have their origin in atmospheric discharges in which case they are called external overvoltages, or they are generated internally by connecting or disconnecting the system, or due to the system fault initiation or extinction.**

译文：具体地，电力系统常常遭受起源于大气放电的过电压，这种情况下称为外部过电压；或者通过连接或断开系统在内部产生，或是由于系统故障的产生和消除。

说明：that 引导的定语从句修饰的主体是 overvoltages。而 in which case 引导的定语从句修饰的主体是整个 that 从句，其中 which case 指代 that 从句的具体内容，they 指代 overvoltages。在 or they are generated internally 中，they 也指代 overvoltages，而 generated 后面除了副词 internally 之外还有两个由 or 连接的介宾短语作为状语，分别以 by 和 due to 开头。为了便于理解，该句子可以拆解为：

Concretely, power systems are often subjected to over-voltages; the over-voltages have their origin in atmospheric discharges; in this case they are called *external over-voltages*; or over-voltages are generated internally by connecting or disconnecting the system; or over-voltages are generated internally due to the system fault initiation or extinction.

**2. Hence, with increasing the system's operating voltage a point is reached when the switching over-voltages became the dominant factor in designing the system's insulation up to approximately 300kV.**

译文：因此，随着系统运行电压提高，达到一点，此时操作过电压成为高达 300kV 左右的系统绝缘设计的主要因素。

说明：出现在介词 with 后面的动名词短语 increasing the system's operating voltage 在翻译时可以按照主谓结构来理解，即 the system's operating voltage increasing。由 when 引导的定语从句修饰的主体是 point，在翻译时可以进行适当的语序调整。

**3. The amplitudes are very high, usually in the order of 1000kV and even more into the transmission line; each stroke is then the maximum insulation strength of the overhead line.**

译文：幅度很大，通常进入输电线的在 1000kV 量级甚至更高；每一击都达到架空线的最大绝缘强度。

说明：词组 in the order of 表示"在……量级"。

**4. The rate of voltage rise of such a traveling wave is at its origin directly proportional to the steepness of the lightning current, which may exceed 100kA/μs, as the voltage levels may simply be calculated by the current multiplied by the effective surge impedance of the line.**

译文：这种行波的电压上升率起初直接与雷电电流的陡度（可能会超过 100kA/μs）成

# Chapter 13　Overvoltage and Lightning Shielding

正比，这是因为电压水平可以简单地通过电流与线路的有效电涌阻抗相乘来计算。

说明：at its origin 表示"最初"。be proportional to 表示"与…成比例"。which 引导的定语从句修饰的主体是 steepness，当其前后的内容联系比较紧密时，翻译时可以将这种定语从句的内容放在括号中说明。be calculated by 表示"通过…计算"，而 multiplied by 表示"与…相乘"。

**5. These types of overvoltages are also effective in the LV distribution systems, where they are either produced by the usually current-limiting switches or where they have been transmitted from the HV distribution systems.**

译文：这些种类的过电压在低压配电系统也有影响，在那里它们要么偶尔是由限流开关引起，要么是从高压配电系统传入。

说明：effective 在此表示"有影响的"。either...or... 表示"要么…要么…"。两个由 where 引导的定语从句修饰的主体都是 the LV distribution systems。

**6. A definition adopted by many authors describes a temporary over-voltage (TOV) as an oscillatory phase-to-ground or phase-to-phase over-voltage of relatively long duration at a given location which is undamped or weakly damped in contrast to switching and lightning over-voltages which are usually highly or very highly damped and of short times.**

译文：被很多作者采用的一个定义将暂态过电压描述为：在指定位置，具有相对较长的持续时间，振荡的，无阻尼或弱阻尼的，相对地或相间过电压，与通常强阻尼或很强阻尼且为期较短的操作过电压和雷电过电压相反。

说明：句中出现了两处相同词语省略情况。在 switching and lightning over-voltages 中省略了 switching 后面的 over-voltages，在 highly or very highly damped 中省略了第一个 highly 后面的 damped。词组 in contrast to 表示"与……相反，与相对"。句子的复杂之处在于，phase-to-phase 后面那个 overvoltage 同时带有 5 个不同结构的定语，分别是① oscillatory，② phase-to-ground or phase-to-phase，③ of relative long duration，④ at a given location，⑤ which is undamped or weakly damped。在译成中文时，一定要保证句子通顺，并尽量符合中文表述习惯。

**7. Examples are the neutral displacement in badly designed star-connected voltage transformers, and arcing ground phenomena in systems with insulated neutrals and the resonance phenomena appearing as a result of the open-circuiting of one or two phases in a three-phase system which can be initiated by faulty circuit breakers or broken conductors.**

译文：例如：设计粗糙的星型电压互感器的中性点位移，中性点不接地的系统中的电弧接地现象，三相系统中由断路器故障或导体损坏引起的一相或两相断路导致的谐振现象。

说明：句子的主语很短，而谓语部分 be 动词带了三个并列的宾语，每一个宾语都有各自的定语。这种情况下，如果仍按原来的句子结构翻译，不易保证三个宾语都与主语有正确的连接关系。因此，可以把 Examples are 译为"例如："，可以涵盖三个宾语。

8. The arc extinguishing ability of the tube-type arrester is related to the size of the power frequency continuous current, which is mostly used for lightning protection on power supply lines.

译文：管型避雷器的灭弧能力与工频续流的大小有关，大多用在供电线路上作避雷保护。

说明："the power frequency continuous current"是工频续流的意思，句子"which is..."是which引导的非限制性定语从句，起到补充说明的作用。

## ABBREVIATIONS (ABBR.)

1. NEMP        nuclear electromagnetic pulse        核子电磁脉冲
2. NNEMP       non-nuclear electromagnetic pulse    非核电磁脉冲
3. TOV         temporary over-voltage               临时过电压

## SUMMARY OF GLOSSARY

1. 外部过电压  external over-voltages
   lightning over-voltages            雷电过电压
   electromagnetic pulses             电磁脉冲
2. 内部过电压  internal over-voltages
   temporary over-voltages            临时过电压
   switching over-voltages            操作过电压
   steady state over-voltages         稳态过电压
3. 工频  power frequency
   power frequency                    工频，电源频率
   working voltage frequency          工作电压频率
   supply frequency                   供电频率

## EXERCISES

**1. Translate the following words or expressions into Chinese.**

(1) magnetic storm        (2) e.g.              (3) lightning stroke
(4) wave front            (5) coulomb           (6) charge separation
(7) lightning steepness   (8) step voltage      (9) valve arrester

**2. Translate the following words or expressions into English.**

(1) 内能                  (2) 阻尼              (3) 雷电防护
(4) 谐振                  (5) 尖端              (6) 操作过电压
(7) 避雷针                (8) 静电的            (9) 使用

# Chapter 13    Overvoltage and Lightning Shielding

**3. Fill in the blanks with proper words or expressions.**

(1) The over-voltages can be first divided into _____ over-voltages and _____ over-voltages according to the locations of sources. The former can be further divided into _____ over-voltages and _____ over-voltages.

(2) According to their duration, which ranges from a few hundred microseconds to several seconds, three main categories of internal over-voltages can be defined: ① _____ over-voltages, if they are oscillatory of power frequency or harmonics, ② _____ over-voltages, if they are heavily damped and of short duration, and ③ _____ over-voltages, if they are of long duration.

(3) The main types of arresters are _____ , _____ and _____ , etc.

# Word-Building (13)    non-; dis-    非，不

## 1. non-[ 前缀 ] 表示：非，无，不

| | | | |
|---|---|---|---|
| conducting | nonconducting | adj. | 不传导的 |
| continuous | noncontinuous | adj. | 不连续的，间断的 |
| renewable | nonrenewable | adj. | 不可再生的，不可更新的 |
| sinusoidal | nonsinusoidal | adj. | 非正弦的 |
| cooperation | non-cooperation | n. | 不合作 |
| essential | non-essential | adj. | 不重要的，非本质的 |
| existent | non-existent | adj. | 不存在的 |
| finite | non-finite | adj. | 非定形的 |
| linear | non-linear | adj. | 非线性的 |
| nuclear | non-nuclear | adj. | 非核的，不涉及原子能的 |

## 2. dis-[ 前缀 ] 表示：否定，相反，不

| | | | |
|---|---|---|---|
| able | disable | vt. | 使残废，使失去能力，丧失能力 |
| advantage | disadvantage | vt. n. | （使）不利，缺点，劣势 |
| agree | disagree | vi. | 不同意 |
| appear | disappear | vi | 消失，不见 |
| charge | discharge | v. n. | 放电；卸载 |
| connect | disconnect | v. | 断开，拆开，分离 |
| like | dislike | vt. | 讨厌，不喜欢 |
| order | disorder | vt. n. | 扰乱（状态），（使）失调，（使）紊乱 |
| place | displace | vt. | 移置，转移 |

# Unit 4

# Review

## I. Summary of Glossaries: Auxiliary Devices and Overvoltages

1. 辅助设备　auxiliary devices
    - instrument　　　　　　　　　　　　　　　　仪表，器械
    - meter　　　　　　　　　　　　　　　　　　仪表，表计
    - transducer　　　　　　　　　　　　　　　　传感器，变换器
    - transformer　　　　　　　　　　　　　　　互感器
    - relay　　　　　　　　　　　　　　　　　　继电器
2. 变压器与互感器　transformers
    - current transformer　　　　　　　　　　　电流互感器
    - voltage transformer　　　　　　　　　　　电压互感器
    - power transformer　　　　　　　　　　　　电力变压器
3. 继电器种类　relay classes
    - magnitude relay　　　　　　　　　　　　　幅度继电器
    - directional relay　　　　　　　　　　　　　定向继电器，极化继电器
    - ratio relay　　　　　　　　　　　　　　　比率继电器
    - differential relay　　　　　　　　　　　　差动继电器
    - pilot relay　　　　　　　　　　　　　　　控制 [引示，辅助] 继电器
4. 保护与后备　protection and backup
    - main protection　　　　　　　　　　　　　主保护
    - primary protection　　　　　　　　　　　　主保护
    - backup protection　　　　　　　　　　　　后备保护
    - local backup protection　　　　　　　　　本地后备保护
    - breaker-failure protection　　　　　　　　断路器失效后备保护
    - remote backup protection　　　　　　　　远程后备保护
5. 故障位置　fault positions
    - internal fault　　　　　　　　　　　　　　内部故障
    - external fault　　　　　　　　　　　　　　外部故障
    - through fault　　　　　　　　　　　　　　直通故障
6. 工频的表示方法
    - power frequency　　　　　　　　　　　　　工频，电源频率
    - working voltage frequency　　　　　　　　工作电压频率
    - supply frequency　　　　　　　　　　　　供电频率
7. 过电压　over-voltages
    - external overvoltage　　　　　　　　　　　外部过电压

| | | |
|---|---|---|
| internal overvoltage | | 内部过电压 |
| lightning overvoltage | | 雷电过电压 |
| switching overvoltage | | 操作过电压 |
| transient overvoltage | | 暂态过电压 |
| temporary overvoltage | | 临时过电压 |

## II. Abbreviations (Abbr.)

1. CT      current transformer      电流互感器
2. VT      voltage transformer      电压互感器
3. BIL      basic insulation levels      基本绝缘水平

# Special Topic(4)  Symbols for Operation

There are many symbols used in the literature on science and technology. Some of them are listed in Table, which are used to represent certain operations such as logical comparisons.

**Table  Familiar Symbols for Operation**

| Symbol | Meaning | Example |
|---|---|---|
| $=$ | Equal to | $\pi = 3.1415\cdots$ |
| $\neq$ | Not equal to | $4.12 \neq 4.13$ |
| $>$ | Greater than | $4.90 > 3.17$ |
| $\gg$ | Much greater than | $980 \gg 12$ |
| $<$ | Less than | $330 < 370$ |
| $\ll$ | Much less than | $0.01 \ll 10$ |
| $\geq$ | Greater than or equal to | $x \geq 3$ is satisfied for $x > 3$ or $x = 3$ |
| $\leq$ | Less than or equal to | $x \leq 5$ is satisfied for $x < 5$ or $x = 5$ |
| $\approx$ | Approximately equal to | $3.14159 \approx 3.14$ |
| $\equiv$ | By definition, establishes a relationship between two or more quantities | |
| $|\ |$ | Absolute magnitude of | $|4| = 4$, where $a = -4$ or $a = 4$ |
| $\Sigma$ | Sum of | $\Sigma(4 + 6 + 8) = 18$ |
| $\because$ | Because | $\because x^2 = 4$ |
| $\therefore$ | Therefore | $\therefore x = 2$ or $x = -2$ |

# Knowledge & Skills (4)  The Subjunctive Mood

## 1. 虚拟语态的概念

The subjunctive mood: a mood of a verb used in some languages for contingent or hypothetical action, an action viewed subjectively, or grammatically subordinate statements.

虚拟语态：一种或一组动词形式，这种形式是用来说明被表示的行为或状态不是事实，而是思想上认为是可能发生的，或带着某些情感来看待的事情。

**2. 虚拟语气所表示的内容并不一定要发生，往往只是"虚拟"和假设。**

例 1：Back-up protection <u>as the name implies</u> is a completely separate arrangement that operates to remove the faulty part should be main protection fail to operate.

译文：顾名思义，后备保护是完全的独立安排，一旦主保护未能动作时由它动作去切除故障部分。

说明：以 should be 为主要表现形式，构造了虚拟语态的句子。should be 相当于 if（如果），provided（倘若）或 in case（万一，一旦）。

例 2：Should the wave encounter a line insulator, the latter will be briefly subjected to a violent overvoltage.

译文：假如该行波遇到线路绝缘子，后者会暂时遭受一个强大的过电压。

说明：以 should 为引导，构造了虚拟语态的句子。should 相当于 if（如果），provided（倘若）或 in case（万一，一旦）。

# Chapter 14
# Faults on Power System

## Part 1  Faults and Their Damage

Faults of many types and causes may appear in electric power systems. Strictly speaking, a fault is any abnormal state of the system, so faults in general consist of short circuits as well as open circuits. We will limit our discussion here to faults that are short circuits. Open circuit faults are much more common than short circuits, and often they are transformed into short circuits by subsequent events. In terms of the seriousness of the consequences of a fault, short circuits are of far greater concern than open circuits, although some open circuits may present a potential hazard to personnel.*

If short circuits are allowed to persist on a power system for an extended period, many or all of the following undesirable effects are likely to occur:

(1) Reduced stability margins for the power system.

(2) Damage to the equipment that is in the vicinity of the fault due to heavy currents, unbalanced currents, or low voltages produced by the short circuit.

(3) Explosions may occur in equipment containing insulating oil during a short circuit which may cause fire resulting in a serious hazard to personnel and damage to other equipment.*

(4) Disruption in the entire power system service area by a succession of protective actions taken by different protection systems, an occurrence known as cascading.*

Which one of these effects will predominate in a given case depends upon the nature and operating conditions of the power system.

## Part 2  Distribution System Faults

On the distribution system, the fault rate is significantly higher compared with the transmission system. With the considerably greater route length of the distribution system, it is not surprising that there are significantly more faults occurring at this level of the system. It is estimated that there may be 25000 faults per annum on the HV distribution system, 70 percent of which involve a connection to Earth. Typically, earth fault current magnitudes are 25kA at 132kV, 1 ~ 2kA at 33kV, and 11kV substations. Below 66kV, the system is designed to restrict earth fault current magnitudes.

Although fault current magnitudes on the distribution system are generally much lower compared with the transmission system, fault clearance times are considerably longer. As a general rule, the fault clearance time will increase the lower the distribution voltage level.* On 11kV systems employing some protection, fault clearance times may exceed 1s. It is important to establish accurate fault clearance times because the tolerable level of current passing through the human body depends on the shock exposure time, which is normally assumed equal to the fault duration. The longer the fault, the lower the tolerable body limit.

The 132kV shielded double circuit system in the UK originally functioned as a transmission grid. Over some time, it evolved into a distribution system as generation displaced to the Super Grid at 275kV and 400kV. These days, new circuits at 132kV are often constructed using unearthed wood pole single circuits, and the vast majority of overhead line construction at 66kV and below has always been unearthed. Hence, distribution substations connected to overhead line circuits may not always have the benefit of an extended earthing system.

On the other hand, most distribution circuits in urban areas will consist of underground cables. The metallic sheaths of these cables form an extended earthing system by virtue of their connection to earthing points along the circuit. This system can be highly interconnected and cover a large area providing low earth impedance values. Also, similar to the action of the shielding wire on overhead lines, the sheath of a cable provides an alternative path for the earth fault current to return to the source. The mutual coupling between the core and sheath of a cable is considerably greater than that between a phase conductor and earth wire of an overhead line, and therefore this mutual effect is much greater for cables.

Older underground cables were constructed with lead sheaths and covered by an insulating layer of bitumen-impregnated hessian (PILSSWA, paper-insulated lead sheathed steel wire armored). It has been found that, over a period of time, this type of insulation degrades leaving the lead sheath of the cable effectively in direct contact with soil. Such cables, therefore, provide an additional fortuitous earth connection. These PILSSWA cables are steadily being replaced by plastic-insulated cables, which will result in a general increase in earth impedances for such urban systems.*

## Part 3  Types of Faults

A fault in a circuit is any failure which interferes with the normal flow of current.

Line-to-line faults now involving ground are less common. Overhead lines are sometimes accidentally brought together by the action of wind, sleet, trees, cranes, airplanes, or damage to supporting structures. Contamination on insulators sometimes results in flashover even during normal voltage conditions.

Permanent faults are caused by lines being on the ground, by insulator strings breaking because of ice loads, by permanent damage to towers, and by lightning-arrester

# Chapter 14  Faults on Power System

failures. Most faults on transmission lines of 115kV and higher are caused by lightning, which results in the flashover of insulators. The high voltage between a conductor and the grounded supporting tower causes ionization, which provides a path to ground for the charge induced by the lightning stroke. Once the ionized path to the ground is established, the resultant low impedance to ground allows the flow of power current from the conductor to the ground and through the ground to the grounded neutral of a transformer or generator, thus completing the circuit.*

Experience has shown that between 70% and 80% of transmission-line faults are single line-to-ground faults, which arise from the flashover of only one line to the tower and ground. The smallest number of faults, roughly 5%, involve all three phases and are called three-phase faults. Other types of transmission-line faults are line-to-line faults, which do not involve ground, and double line-to-ground faults. All the above faults except the three-phase type are unsymmetrical and cause an imbalance between the phases.

The conductors of underground cables are separated from each other and ground by solid insulation. These materials undergo some deterioration with age, particularly if overloads on the cables have resulted in their operation at elevated temperatures. Deterioration of the insulation may result in the failure of the material to retain its insulating properties, and short circuits will develop between the cable conductors. The possibility of cable failure is increased if lightning or switching produces a transient voltage of abnormally high values between the conductors.

Transformer failures may be the result of insulation deterioration combined with over-voltages due to lightning or switching transients. Short circuits due to insulation failure between, adjacent turns of the same winding may result from suddenly applied over-voltages. Major insulation may fail, permitting arcs to be established between primary and secondary windings or between winding and grounded metal parts such as the core or tank.

Generators may fail due to the breakdown of the insulation between adjacent turns in the same slot, resulting in a short circuit in a single turn of the generator. Insulation breakdown may also occur between one of the windings and the grounded steel structure in which the coils are embedded. A breakdown between different windings lying in the same slot results in short-circuiting extensive sections of the machine.

## Part 4  Calculation of Fault Currents

The bus impedance matrix is preferred by some engineers for load-flow studies but finds its greatest value in fault calculations.*

Fault calculations consist of determining these currents for various types of faults at various locations in the system. The data obtained from fault calculations also serve to determine the settings of relays that control the circuit breakers. Analysis by symmetrical components is a powerful tool that makes the calculation of unsymmetrical faults almost as easy as the calculation of three-phase faults.

Unbalanced voltages and currents can be resolved into their symmetrical components. Problems are solved by treating each set of components separately and superimposing the results.

In balanced networks having no coupling between phases, the currents of one phase sequence induce voltage drops of like sequence only. Impedances of circuit elements to currents of different sequences are not necessarily equal.

A knowledge of the positive-sequence network is necessary for fault calculations. If the fault calculations involve unsymmetrical faults on otherwise symmetrical systems, the negative and zero-sequence networks are needed also. Synthesis of the zero-sequence network requires particular care, because the zero-sequence network may differ from the others considerably.

Again it is the digital computer that is invaluable in making fault calculations.

## Part 5   Faults Isolation and Recovery

Although the occurrence of short circuits is somewhat of a rare event, it is of utmost importance that steps be taken to remove the short circuits from a power system as quickly as possible.* In modern power systems, this short circuit removal process is executed automatically, that is, without human intervention. The equipment that does this job is known collectively as the protection system.

Faults can be very destructive to power systems. A great deal of study, development of devices, and design of protection schemes have resulted in continual improvement in the prevention of damage to transmission lines and equipment and interruptions in generation following the occurrence of a fault.

Faults caused by surges are usually of such short duration that any circuit breakers which may open will reclose automatically after a few cycles to restore normal operation. If arresters are not involved or faults are permanent the faulted sections of the system must be isolated to maintain normal operation of the rest of the system.

The operation of circuit breakers is controlled by relays which sense the fault. In the application of relays zones of protection are specified to define the parts of the system for which various relays are responsible.* One relay will also back up another relay if an adjacent zone fails to respond.

Opening circuit breakers to isolate the faulted portion of the line from the rest of the system interrupt the flow of current in the ionized path and allows deionization to take place. After an interval of about 20 cycles to allow deionization, breakers can usually be reclosed without reestablishing the arc. Experience in the operation of transmission lines has shown that ultra-high-speed reclosing breakers successfully reclose after most faults. Of those cases where reclosure is

# Chapter 14　Faults on Power System

not successful, an appreciable number are caused by permanent faults where reclosure would be impossible regardless of the interval between opening and reclosing.

The current that flows in different parts of a power system immediately after the occurrence of a fault differs from that flowing a few cycles later just before circuit breakers are called upon to open the line on both sides of the fault, and both these currents differ widely from the current which would flow under steady-state conditions if the fault were not isolated from the rest of the system by the operation of circuit breakers.

## NEW WORDS AND EXPRESSIONS

| 1. | stability margin | | 稳定边际 [储备，裕度] |
| --- | --- | --- | --- |
| 2. | vicinity | n. | 邻近，附近 |
| 3. | a succession of | | 一连串 |
| 4. | cascading | n. | 级联，层叠 |
| 5. | tolerable body limit | | 人体承受极限 |
| 6. | double circuit | | 双回路 |
| 7. | unearthed | adj. | 未接地的 |
| 8. | sheath | n. | 套管，包皮 |
| 9. | lead sheath | | 铅皮，铅防护套 |
| 10. | bitumen | n. | 沥青 |
| 11. | hessian | n. | 粗麻布 |
| 12. | armored | adj. | 披甲的，装甲的 |
| 13. | fortuitous | adj. | 偶然的，幸运的 |
| 14. | sleet | n. | 冰雪，雨夹雪 |
| 15. | crane | n. | 起重机 |
| 16. | supporting structure | | 支承结构，固定架 |
| 17. | string | n. | 一串，一行 |
| 18. | resultant | n. | 作为结果而发生的，合成的 |
| 19. | unsymmetrical | adj. | 非对称的，不均匀的 |
| 20. | symmetrical component | | 对称分量 |
| 21. | resolved into | | 使分解为 |
| 22. | superimpose | v. | 叠加 |
| 23. | synthesis | n. | 综合，合成 |
| 24. | utmost | adj. | 极度的 |
| 25. | deionization | n. | 消电离作用 |
| 26. | appreciable | adj. | 可感知的，可评估的 |

## PHRASES

**1. evolve into: 发展 [进化] 成，演变为**

Example: Over some time, it evolved into a distribution system as generation displaced to the Super Grid at 275kV and 400kV.

经过一段时间，由于发电转移到 275～400kV 的超级电网，它演变成为一个配电系统。

**2. by virtue of: 依靠，由于**

Example: The metallic sheaths of these cables form an extended earthing system by virtue of their connection to earthing points along the circuit.

依靠与电路沿线的接地点的连接，这些电缆的金属包皮形成一个扩展的接地系统。

**3. arise from: 起于，由…出身**

Example: Experience has shown that between 70% and 80% of transmission-line faults are single line-to-ground faults, which arise from the flashover of only one line to the tower and ground.

实践表明，70%～80%的输电线故障都是单相接地短路故障，源自单相线路对杆塔和大地的闪络。

**4. make...as easy as: 使…和…一样简单**

Example: Analysis by symmetrical components is a powerful tool that makes the calculation of unsymmetrical faults almost as easy as the calculation of three-phase faults.

用对称分量分析是一种很有效的方法，它使不对称故障分析几乎和三相故障分析一样简单。

**5. regardless of: 不管，不顾**

Example: Of those cases where reclosure is not successful, an appreciable number are caused by permanent faults where reclosure would be impossible regardless of the interval between opening and reclosing.

在那些重合闸没成功的案例中，有数量可观的是由永久故障引起的，在那里不管断开和重合的间隔长短都不可能重合。

## COMPLICATED SENTENCES

**1. In terms of the seriousness of the consequences of a fault, short circuits are of far greater concern than open circuits, although some open circuits may present a potential hazard to personnel.**

译文：尽管某些断路会给人们带来潜在危害，但根据故障后果的严重性，短路远远比断路更值得关注。

说明：In terms of 表示"根据，按照，在……方面"。far more than 表示"远胜于"，其中 more 代指形容词的比较级形式。of far greater concern than 表示"远比…重要，远比…值得关注"。

**2. Explosions may occur in equipment containing insulating oil during a short circuit which may cause fire resulting in a serious hazard to personnel and damage to other equipment.**

译文：含有绝缘油的设备在短路时会引发爆炸和火灾，这会对人们造成严重危害，也

# Chapter 14　Faults on Power System

会破坏其他设备。

说明：名词短语中，当短语主体名词所带的定语较长、较多或较为复杂时，可以转为主谓结构的句子进行翻译，某些或全部定语不再当作定语，而作为谓语使用。为便于理解，上述句子可以分别拆解为下述两种形式：

Explosions may occur in equipment containing insulating oil during a short circuit and may cause fire resulting in a serious hazard to personnel and damage to other equipment.

Explosions, which may occur in equipment containing insulating oil during a short circuit, may cause fire resulting in a serious hazard to personnel and damage to other equipment.

**3. Disruption in the entire power system service area by a succession of protective actions taken by different protection systems, an occurrence known as cascading.**

译文：由不同保护系统采取的一连串保护动作引起整个电力系统供电区的断电，即大家知道的级联现象。

说明：词组 a succession of 表示"一连串"。known as 表示"作为……被人所知"，可以译为"大家知道的，众所周知的"。an occurrence 是 disruption 的同位语。

**4. As a general rule, the fault clearance time will increase the lower the distribution voltage level.**

译文：通常，配电电压水平越低，故障切除时间越长。

说明：词组 as a rule 表示"通常，一般"。这个句子是句型 the more1... the more2... 的变形，其中 more1 和 more2 代指形容词或副词的比较级。

**5. These PILSSWA cables are steadily being replaced by plastic-insulated cables, which will result in a general increase in earth impedances for such urban systems.**

译文：这些 PILSSWA 电缆被塑料绝缘电缆逐步替换，能够全面提高城市系统对地阻抗。

说明：steadily 表示"稳定地，有规则地"。replaced by 表示"取代，以…代替"。a general increase 意思是全面提高。

**6. Once the ionized path to ground is established, the resultant low impedance to the ground allows the flow of power current from the conductor to the ground and through the ground to the grounded neutral of a transformer or generator, thus completing the circuit.**

译文：接地电离通道一旦建立，其对应的对地低阻抗会使电流从导线流入大地，并经过大地流入变压器或者发电机的接地中线，从而形成一个完整回路。

**7. The bus impedance matrix is preferred by some engineers for load-flow studies but finds its greatest value in fault calculations.**

译文：一些工程师喜欢用结点阻抗矩阵研究电力潮流，但是阻抗矩阵的最大价值在于故障计算。或者：结点阻抗矩阵被一些工程师首选用在潮流研究上，但却在故障计算中找到了其最大价值。

说明：Bus impedance matrix 是整个句子的主语，is... 和 finds 分别做表语和谓语。

**8. Although the occurrence of short circuits is somewhat of a rare event, it is of utmost importance that steps be taken to remove the short circuits from a power system as quickly as possible.**

译文：尽管发生短路是相当罕见的事，尽快采取措施将短路故障从电力系统切除仍是极为重要的。

说明：somewhat of 表示"相当地"，近似于副词 quite。somewhat of a rare event 也可写为 a somewhat of rare event，意思为"相当罕见的事件"。of utmost importance 意思是"极为重要，具有极为重要的意义"。it is of utmost importance that steps be taken 是强调句，句型 it is A that B，强调 B 的属性 A。steps be taken 比 steps should be taken 语气更强烈一些。

**9. In the application of relays zones of protection are specified to define the parts of the system for which various relays are responsible.**

译文：在继电器应用中，指定保护区域来定义各继电器负责的系统部分。

说明：整个句子主语是 zones of protection。翻译时一定不能把 relays zones 看成是一体的。for which 以介词加连词的形式引导一个定语从句，修饰的主体是 parts。

## ABBREVIATIONS (ABBR.)

1. PILSS-WA    paper-insulated lead sheathed steel wire armored    纸绝缘铅护套钢丝铠装
2. UK    United Kingdom    英国，联合王国

## SUMMARY OF GLOSSARY

1. 回路系统    circuit systems
   - single circuit system    单回路系统
   - double circuit system    双回路系统
2. 故障分类（1） types of faults
   - short circuit    短路
   - open circuit    断路
3. 故障分类（2） types of faults
   - permanent fault    永久故障
   - temporary fault    临时故障，短时故障
4. 对称分量    symmetrical components
   - positive-sequence    正序
   - negative-sequence    负序
   - zero-sequence    零序

# Chapter 14  Faults on Power System

## EXERCISES

**1. Translate the following words or expressions into Chinese.**

(1) cascading  (2) unearthed  (3) lead sheath
(4) imbalance  (5) recovery  (6) intervention

**2. Translate the following words or expressions into English.**

(1) 稳定裕度  (2) 双回路  (3) 绝缘子串
(4) 叠加  (5) 对称分量  (6) 非对称的

**3. Fill in the blanks with proper words or expressions.**

(1) Strictly speaking, a fault is any _____ state of the system, so faults in general consist of _____ as well as _____ .

(2) As a general rule, the fault clearance time will _____ the lower the distribution voltage level.

(3) Analysis by _____ is a powerful tool that makes the calculation of unsymmetrical faults almost as easy as the calculation of three-phase faults.

# Word-Building (14)  magni-，meg(a)-，micro-  巨大，微小

## 1. magni-  表示：大，长

| | | |
|---|---|---|
| magnific | adj. | 庄严的，崇高的 |
| magnification | n. | 扩大，放大倍率 |
| magnificent | adj. | 华丽的，高尚的，宏伟的 |
| magnify | vt. | 放大，扩大，赞美，夸大，夸张 |
| magnifier | n. | 放大镜，放大器 |
| magnitude | n. | 大小，幅度；数量，巨大，量级 |

## 2. meg(a)-  表示：大，巨大；meglo-  表示：巨大，扩大

| | | |
|---|---|---|
| megaphone | n. | 扩音器，喇叭 |
| megadebt | n. | 巨额债务 |
| megacity | n. | （人口超过100万的）大城市 |
| megalopolis | n. | 巨大都市，人口稠密地带 |
| megalograph | n. | 显微图形放大装置 |

## 3. micro-  表示："极微小，仪器或工具用以扩大者"之义

| | | |
|---|---|---|
| microcircuit | n. | 微电路 |
| microelectronic | n. | 微电子的 |
| microelement | n. | 微型元件，微量元素 |
| micromotor | n. | 微型马达 |

| | | |
|---|---|---|
| microwave | *n.* | 微波 |
| microcomputer | *n.* | 微机,微型(电子)计 |
| microcosmic | *adj.* | 微观世界的,微观的 |
| microeconomic | *n.* | 微观经济 |
| microanalysis | *n.* | 微量分析 |
| microfilm | *n.* | 缩微胶片 |
| microphone | *n.* | 扩音器,麦克风 |
| microscope | *n.* | 显微镜 |

# Chapter 15
# Stability of Power System

## Part 1  Basic Concepts and Definitions

*Power system stability* may be broadly defined as that property of a power system that enables it to remain in a state of operating equilibrium under normal operating conditions and to regain an acceptable state of equilibrium after being subjected to a disturbance.*

Instability in a power system may be manifested in many different ways depending on the system configuration and operating mode. Traditionally, the stability problem has been one of maintaining synchronous operation. Since power systems rely on synchronous machines for the generation of electrical power, a necessary condition for satisfactory system operation is that all synchronous machines remain in synchronism or, colloquially, " in step " .* This aspect of stability is influenced by the dynamics of generator rotor angles and power-angle relationships.

Instability may also be encountered without loss of synchronism. For example, a system consisting of a synchronous generator feeding and an induction motor load through a transmission line can become unstable because of the collapse of load voltage. Maintenance of synchronism is not an issue in this instance; instead, the concern is stability and control of voltage. This form of instability can also occur in loads covering an extensive area supplied by a large system.

In the evaluation of stability, the concern is the behavior of the power system when subjected to a transient disturbance. The disturbance may be small or large. Small disturbances in the form of load changes take place continually, and the system adjusts itself to the changing conditions. The system must be able to operate satisfactorily under these conditions and successfully supply the maximum amount of load. It must also be capable of surviving numerous disturbances of a severe nature, such as a shortcircuit on a transmission line, loss of a large generator or load, or loss of a tie between two subsystems. The system response to a disturbance involves much of the equipment. For example, a short-circuit on a critical element followed by its isolation by protective relays will cause variations in power transfers, machine rotor speeds, and bus voltages, the voltage variations will actuate both generator and transmission system voltage regulators, the speed variations will actuate prime mover governors, the change in tie line loading may actuate generation controls, the changes in voltage and frequency will affect loads on the system in varying degrees depending on their characteristics. In addition, devices used to protect individual equipment may respond to variations in system variables and thus affect the system's performance. In any given situation, however, the responses of only a limited amount of equipment may be significant. Therefore, many assumptions are usually made to simplify the problem and to focus on factors influencing the specific type of stability problem. The understanding of stability problems is greatly facilitated by the classification of stability into various categories.

# Part 2　Classification of Stability

Power system stability is a single problem; however, it is impractical to study it as such. As seen in the previous section, the instability of a power system can take different forms and can be influenced by a wide range of factors. Analysis of stability problems, identification of essential factors that contribute to instability, and formation of methods of improving stable operation are greatly facilitated by the classification of stability into appropriate categories. These are based on the following considerations:

- The physical nature of the resulting instability;
- The size of the disturbance considered;
- The devices, processes, and time that must be taken into consideration in order to determine stability;
- The most appropriate method of calculation and prediction of stability.

Figure 15.1 gives an overall picture of the power system stability problem, identifying its classes and subclasses in terms of the categories. As a practical necessity, the classification has been based on a number of diverse considerations, making it difficult to select distinct categories and provide definitions that are rigorous yet convenient for practical use. For example, there is some overlap between mid-term/long-term stability and voltage stability. With appropriate models for loads, on-load transformer tap changers, and generator reactive power limits, mid-term/long-term stability simulations are ideally suited for dynamic analysis of voltage stability. Similarly, there is an overlap between transient, mid-term, and long-term stability: all three use similar analytical techniques for the simulation of the nonlinear time domain response of the system to large disturbances. Although the three categories are concerned with different aspects of the stability problem, in terms of analysis and simulation they are extensions of one another without clear boundaries.*

While the classification of power system stability is an effective and convenient means to deal with the complexities of the problem, the overall stability of the system should always be kept in mind. Solutions to stability problems of one category should not be at the expense of another. It is essential to look at all aspects of the stability phenomena and each aspect from more than one viewpoint. This requires the development and wise use of different kinds of analytical tools. In this regard, some degree of overlap in the phenomena being analyzed is desirable.

# Chapter 15  Stability of Power System

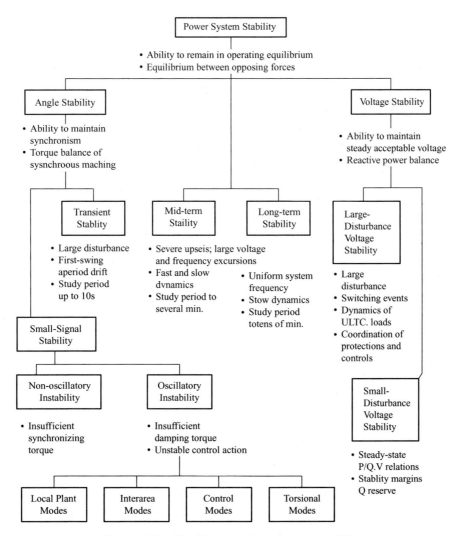

**Figure 15.1**  Classification of power system stability

# Part 3  Angle Stability

*Angle stability*, also called *synchronous stability*, is a condition of equilibrium between opposing forces. The mechanism by which interconnected synchronous machines maintain synchronism with one another is through restoring forces, which act whenever there are forces tending to accelerate or decelerate one or more machines with respect to other machines.* Under steady-state conditions, there is an equilibrium between the input mechanical torque and the output electrical torque of each machine, and the speed remains constant. If the system is perturbed this equilibrium is upset, resulting in acceleration or deceleration of the rotors of the machines according to the laws of motion of a rotating body. If one generator temporarily runs faster than another, the angular position of its rotor relative to that of the slower machine will advance. The resulting angular difference transfers part of the load from the slow machine to the fast machine, depending on the power-angle relationship. This tends to reduce the speed difference and hence the angular separation. The power-angle relationship, as discussed above, is highly nonlinear. Beyond a certain limit, an increase in angular separation is accompanied by a decrease in power transfer; this increases the angular separation further and leads to instability. For any given situation, the stability of the system depends on whether or not the deviations in angular positions of the rotors result in sufficient *restoring torques*.

When a synchronous machine loses synchronism or "falls out of step" with the rest of the system, its rotor runs at a higher or lower speed than that required to generate voltages at system frequency. The "slip" between the rotating stator field (corresponding to system frequency) and the rotor field results in large fluctuations in the machine power output, current, and voltage; this causes the protection system to isolate the unstable machine from the system.

Loss of synchronism can occur between one machine and the rest of the system or between groups of machines. In the latter case, synchronism may be maintained within each group after its separation from the others.

With electric power systems, the change in electrical torque $\Delta T_e$ of a synchronous machine following a perturbation can be resolved into two components: one is the component of torque change in phase with the rotor angle perturbation $\Delta \delta$ and is referred to as the *synchronizing torque* component; the other is the component of torque in phase with the speed deviation $\Delta \omega$ and is referred to as the *damping torque* component.

$$\Delta T_e = T_S \Delta \delta + T_D \Delta \omega$$

Where $T_S$ and $T_D$ are the synchronizing and damping torque coefficients respectively.

System stability depends on the existence of both components of torque for each of the synchronous machines. Lack of sufficient synchronizing torque results in instability through an aperiodic drift in rotor angle. On the other hand, a lack of sufficient damping torque results in *oscillatory instability*.

For convenience in analysis and for gaining useful insight into the nature of stability problems, it is usual to characterize the rotor angle stability phenomena in terms of the following two categories:

(1) *Small-signal (or small-disturbance) stability* is the ability of the power system to maintain synchronism under small disturbances. Such disturbances occur continually in the system because of small variations in loads and generation. The disturbances are considered sufficiently small for linearization of the system equations to be permissible for purposes of analysis. The nature of the system response to small disturbances depends on a number of factors including the initial operation, the transmission system strength, and the type of generator excitation controls used.

(2) *Transient stability* is the ability of the power system to maintain synchronism when subjected to a severe transient disturbance. The resulting system response involves large ex-

# Chapter 15   Stability of Power System

cursions of generator rotor angles and is influenced by the nonlinear power-angle relationship. Stability depends on both the initial operating state of the system and the severity of the disturbance. Usually, the system is altered so that the post-disturbance steady-state operation differs from that prior to the disturbance.

The term *dynamic stability* has also been widely used in the literature as a class of rotor angle stability. However, it has been used to denote different aspects of the phenomenon by different authors. In North American literature, it has been used mostly to denote small-signal stability in the presence of automatic control devices (primarily generator voltage regulators) as distinct from the classical steady-state stability without automatic controls.* In the French and German literature, it has been used to denote what we have termed here *transient stability*. Since much confusion has resulted from the use of the term *dynamic stability*, both CIGRE and IEEE have recommended that it not be used.

There are two factors that can act as guideline criteria for the relative stability of a generating unit within a power system. These are the angular swing of the machine during and following fault conditions and the *critical clearing time*. *Equal-area criterion* is applied to fault clearing when power is transmitted during the fault.

## Part 4   Voltage Stability and Collapse

*Voltage stability* is the ability of a power system to maintain steady acceptable voltage at all busses in the system under normal operating conditions and after being subjected to a disturbance. A system enters a state of voltage instability when a disturbance, increase in load demand, or change in system condition causes a progressive and uncontrollable drop in voltage. The main factor causing instability is the inability of the power system to meet the demand for reactive power. The heart of the problem is usually the voltage drop that occurs when active power and reactive power flow through inductive reactances associated with the transmission network.

A criterion for voltage stability is that, at a given operating condition for every bus in the system, the bus voltage magnitude increases as the reactive power injection at the same bus is increased. A system is voltage unstable if, for at least one bus in the system, the bus voltage magnitude ($V$) decreases as the reactive power injection ($Q$) at the same bus is increased. In other words, a system is voltage stable if $V$-$Q$ sensitivity is positive for every bus and voltage unstable if $V$-$Q$ sensitivity is negative for at least one bus.

A progressive drop in bus voltage can also be associated with rotor angles going out of step. For example, the gradual loss of synchronism of machines as rotor angles between two groups of machines approach or exceed 180° would result in very low voltages at intermediate points in the network. In contrast, the type of sustained fall of voltage that is related to voltage instability occurs where rotor angle stability is not an issue.

Voltage instability is essentially a local phenomenon; however, its consequences may have a widespread impact. *Voltage collapse* is more complex than simple voltage instability and is usually the result of a sequence of events accompanying voltage instability leading to a low-voltage profile in a significant part of the power system.

Voltage instability may occur in several different ways.

In complex practical power systems, many factors contribute to the process of system collapse because of voltage instability: strength of the transmission system; power-transfer levels; load characteristics; generator reactive power capability limits; and characteristics of reactive power compensating devices. In some cases, the problem is compounded by the uncoordinated action of various controls and protective systems.

For purposes of analysis, it is useful to classify voltage stability into the following two subclasses:

(1) *Large-disturbance voltage stability* is concerned with a system's ability to control voltages following large disturbances such as system faults, loss of generation, or circuit contingencies. This ability is determined by the system-load characteristics and the interactions of both continuous and discrete controls and protections.

(2) *Small-disturbance voltage stability* is concerned with a system's ability to control voltages following small perturbations such as incremental changes in system load. This form of stability is determined by the characteristics of the load, continuous controls, and discrete controls at *a given instant of time*. This concept is useful in determining, at any instant, how the system voltage will respond to small system changes.

Voltage instability does not always occur in its pure form. Often the angle and voltage instabilities go hand in hand. One may lead to the other and the distinction may not be clear. However, a distinction between angle stability and voltage stability is important for understanding the underlying causes of the problems in order to develop appropriate procedures.

## NEW WORDS AND EXPRESSIONS

| 1. | equilibrium | n. | 平衡，均衡 |
| 2. | regain | v. | 重得，重新达到 |
| 3. | disturbance | n. | 干扰，扰动 |
| 4. | instability | n. | 不稳定（性） |
| 5. | manifest | vt. | 出现，表现 |
| 6. | synchronism | n. | 同步性 |
| 7. | colloquially | adv. | 用通俗语 |
| 8. | dynamics | n. | 动力学 |
| 9. | collapse | n. | 崩溃 |
| 10. | issue | n. | 重点，难题 |
| 11. | surviving | adj. | 幸免于，从困境中挺过来 |
| 12. | tie | n. | 联络线，连接线 |
| 13. | assumption | n. | 假定，设想 |
| 14. | formation | n. | 形成，构成 |

# Chapter 15  Stability of Power System

| | | | |
|---|---|---|---|
| 15. | time span | n. | 时间间隔 |
| 16. | rigorous | adj. | 严格的，严厉的 |
| 17. | overlap | v. | 交迭 |
| 18. | simulation | n. | 仿真，模拟 |
| 19. | analytical | adj. | 分析的，解析的 |
| 20. | opposing | adj. | 反向的，相反的 |
| 21. | perturb | v. | 扰乱，干扰，使混乱 |
| 22. | upset | v. | 颠覆，推翻，扰乱 |
| 23. | deviation | n. | 背离 |
| 24. | fluctuation | n. | 波动，起伏 |
| 25. | coefficient | n. | 系数 |
| 26. | linearization | n. | 线性化 |
| 27. | permissible | adj. | 可允许的 |
| 28. | excursion | n. | 偏移，漂移 |
| 29. | post-disturbance | | 扰动后的 |
| 30. | prior to | | 在前，居先 |
| 31. | progressive | adj. | 逐渐的，渐进的，发展的 |
| 32. | profile | n. | 轮廓，断面 |
| 33. | incremental change | | 递增，递增量 |

## PHRASES

**1. in any given situation: 在任何特定情况下**

Example: In any given situation, however, the responses of only a limited amount of equipment may be significant.

然而，在任何特定情况下，只有有限数量的设备响应会比较重要。

**2. focus on: 集中于，把注意力落在……上**

Example: Therefore, many assumptions are usually made to simplify the problem and to focus on factors influencing the specific type of stability problem.

因此，通常要做很多假设，以把注意力集中在影响特定类型稳定性问题的因素上。

**3. at the expense of: 在损害…的情况下，以牺牲…为代价**

Example: Solutions to stability problems of one category should not be at the expense of another.

某一类稳定问题的解决不应以牺牲其他类别为代价。

**4. in this regard: 不管，不顾**

Example: In this regard, some degree of overlap in the phenomena being analyzed is desirable.

在这点上，被分析的三种现象之间的某种程度的重叠，实际上是值得要的。

## COMPLICATED SENTENCES

**1. Power system stability may be broadly defined as that property of a power system that enables it to remain in a state of operating equilibrium under normal operating conditions and to regain an acceptable state of equilibrium after being subjected to a disturbance.**

译文：电力系统稳定性可以广泛定义为电力系统的一个特性，即在正常工作条件下保持运行平衡状态，在受到干扰以后可以重新回到可接受的平衡状态。

说明：在类似定义等表述语句中，往往说明的主体带有很长的定语，而且定语的结构可能还比较复杂。在翻译时，为了避免拗口，可以将定语部分转换为单独的句子来翻译，只要在原来的主体和定语部分之间建立适当的联系即可。

**2. Since power systems rely on synchronous machines for the generation of electrical power, a necessary condition for satisfactory system operation is that all synchronous machines remain in synchronism or, colloquially, "in step".**

译文：由于电力系统依靠同步电机发电，系统正常运行的一个必要条件就是所有的同步电机保持同步，通俗一点讲就是"步调一致"。

说明：in synchronism 和 in step 都可以表示"同步"，前者为比较专业的用法，后者通俗一些。

**3. Although the three categories are concerned with different aspects of the stability problem, in terms of analysis and simulation they are extensions of one another without clear boundaries.**

译文：尽管在分析和仿真中这三个类别涉及稳定性问题的不同方面，它们之间的确没有明确的界限。

说明：词组 be concerned with 原意是"干预，参与"，在此表示"涉及，关注"。in terms of 表示"在……方面"。one another 意思是"彼此之间"，修饰后面的 without 短语。

**4. The mechanism by which interconnected synchronous machines maintain synchronism with one another is through restoring forces, which act whenever there are forces tending to accelerate or decelerate one or more machines with respect to other machines.**

译文：相联的同步电机彼此保持同步的机理是利用回复力，当有一台或多台电机在外力作用下相对于其他电机加速或减速运转时，回复力起作用。

说明：句子的主干是 The mechanism is through restoring forces。其中主语 mechanism 带有一个由 by which 引导的定语从句，而 restoring force 带有一个由 which 引导的定语从句，该定语从句中还含有一个由 whenever 引导的时间状语从句。词组 with respect to 在此表示"相对于……"。

# Chapter 15  Stability of Power System

**5. In North American literature, it has been used mostly to denote small-signal stability in the presence of automatic control devices (primarily generator voltage regulators) as distinct from the classical steady-state stability without automatic controls.**

译文：在北美的文献中，它主要用于表示在有自动控制装置（主要是发电机电压调节器）存在情况下的小信号稳定，以区别于无自动控制的经典稳态稳定性。

说明：词组 in the presence of 表示"在……面前，有……存在时"。

## ABBREVIATIONS (ABBR.)

| | | | |
|---|---|---|---|
| 1. | CIGRE | conference international des grands reseaux electriques; international conference on large high voltage electric system | 国际大电网会议 |
| 2. | IEEE | Institute of Electrical and Electronics Engineers | 电气和电子工程师协会 |

## SUMMARY OF GLOSSARY

1. 扰动
   - disturbance —— 扰动
   - fault —— 故障
   - contingency —— 意外事故
   - perturbation —— 动摇，混乱

2. 同步性　synchronism
   - synchronous —— 同步的
   - in synchronism —— 同步，保持同步性
   - in step —— 同步，步调一致
   - out of step —— 失步，失去同步的
   - asynchronous —— 异步的，不同步的
   - nonsynchronous —— 非同步的

3. 稳定性分类（1）　types of stability
   - steady-state stability —— 稳态稳定性
   - dynamic stability —— 动态稳定性
   - transient stability —— 暂态稳定性
   - small-disturbance stability —— 小扰动稳定性

4. 稳定性分类（2）　types of stability
   - synchronous stability —— 同步稳定性
   - angle stability —— 功角稳定性
   - voltage stability —— 电压稳定性

5. 转矩　torques
   - synchronizing torque —— 同步转矩

| | |
|---|---|
| damping torque | 阻尼转矩 |
| restoring torque | 回复力矩，回复转矩 |
| input mechanical torque | 输入机械转矩 |
| output electrical torque | 输出电磁转矩 |

# EXERCISES

**1. Translate the following words or expressions into Chinese.**

(1) rotor angle  (2) equilibrium  (3) equal-area criterion
(4) angular difference  (5) post-disturbance  (6) laws of motion
(7) simulation  (8) fluctuation  (9) angle stability

**2. Translate the following words or expressions into English.**

(1) 临界切除时间  (2) 功角  (3) 电压崩溃
(4) 意外事故  (5) 等面积法则  (6) 回复力矩

**3. Fill in the blanks with proper words or expressions.**

(1) Power system stability may be broadly defined as that property of a power system that enables it to remain in a state of operating _____ under normal operating conditions and to regain an acceptable state of equilibrium after being subjected to a _____.

(2) With electric power systems, the change in electrical torque $\Delta T_e$ of a synchronous machine following a perturbation can be resolved into two components: one is the component of torque change in phase with the _____ perturbation $\Delta\delta$ and is referred to as the _____ torque component; the other is the component of torque in phase with the _____ deviation $\Delta\omega$ and is referred to as the _____ torque component.

(3) Voltage _____ is more complex than simple voltage _____ and is usually the result of a sequence of events accompanying voltage instability leading to a low-voltage profile in a significant part of the power system.

# Word-Building (15)　-ity, -age　-度，-性；总数

**1. -ity**　[名词词尾]，由形容词演变而来，构成抽象名词，表示"…性，…度"

| | | | | | |
|---|---|---|---|---|---|
| durable | adj. | 耐用的 | durability | n. | 耐用性 |
| feasible | adj. | 可行的 | feasibility | n. | 可行性 |
| flexible | adj. | 灵活的 | flexibility | n. | 灵活性 |
| reliable | adj. | 可靠的 | reliability | n. | 可靠性 |

# Chapter 15　Stability of Power System

| secure | *adj.* | 安全的 | security | *n.* | 安全性 |
| stable | *adj.* | 稳定的 | stability | *n.* | 稳定性 |
| sensitive | *adj.* | 灵敏的 | sensitivity | *n.* | 灵敏（度），灵敏性 |

**2. -age**　[名词词尾]，接在数量单位名词后，表示：**数量的总和，集合**。

| mile | *n.* | 英里（里程单位） | mileage | *n.* | 英里数，英里里程 |
| ton | *n.* | 吨（重量单位） | tonnage | *n.* | 吨位，总吨数 |
| volt | *n.* | 伏特（电压单位） | voltage | *n.* | 电压，伏特数 |

# Chapter 16
# Voltage Regulation and Reactive Power Compensation

## Part 1   Production and Absorption of Reactive Power

Bus voltages will change according to load flow variation, whereas they are mainly affected by the production and absorption of reactive power at those buses.

Synchronous generators can generate or absorb reactive power depending on the excitation. When overexcited they supply reactive power, and when underexcited they absorb reactive power. The capability to continuously supply or absorb reactive power is, however, limited by the field current, armature current, and end-region heating limits.

Overhead lines, depending on the load current, either absorb or supply reactive power. At loads below the natural (surge impedance) load, the lines produce net reactive power; at loads above the natural load, the lines absorb reactive power.

Underground cables, owing to their high capacitance, have high natural loads. They are always loaded below their natural loads and hence generate reactive power under all operating conditions.

Transformers always absorb reactive power regardless of their loading; at no load, the shunt magnetizing reactance effects predominate; and at full load, the series leakage inductance effects predominate.

Loads normally absorb reactive power. A typical load bus supplied by a power system is composed of a large number of devices. The composition changes depending on the day, season, and weather conditions. The composite characteristics are normally such that a load bus absorbs reactive power. Both active power and reactive power of the composite loads vary as a function of voltage magnitudes. Loads at low-lagging power factors cause excessive voltage drops in the transmission network and are uneconomical to supply. Industrial consumers are normally charged for reactive as well as active power, this gives them an incentive to improve the load power factor by using shunt capacitors.

Compensating devices are usually added to supply or absorb reactive power and thereby control the reactive power balance in a desired manner.

# Chapter 16  Voltage Regulation and Reactive Power Compensation

## Part 2  Objectives and Methods of Voltage Regulation

For efficient and reliable operation of power systems, the control of voltage and reactive power should satisfy the following objectives:

(1) Voltages at the terminals of all equipment in the system are within acceptable limits. Both utility equipment and customer equipment are designed to operate at a certain voltage rating. Prolonged operation of the equipment at voltages outside the allowable range could adversely affect their performance and possibly cause them damage.

(2) System stability is enhanced to maximize utilization of the transmission system. As described in some academic papers, voltage, and reactive power control have a significant impact on system stability.

(3) The reactive power flow is minimized so as to reduce $I^2R$ and $I^2X$ losses to a practical minimum. This ensures that the transmission system operates efficiently, i.e., mainly for active power transfer.

The problem of maintaining voltages within the required limits is complicated by the fact that the power system supplies power to a vast number of loads and is fed from many generating units.* As loads vary, the reactive power requirements of the transmission system vary. This is abundantly clear from the performance characteristics of transmission lines. Since reactive power cannot be transmitted over long distances, voltage control has to be effected by using special devices dispersed throughout the system. This is in contrast to the control of frequency which depends on the overall system active power balance. The proper selection and coordination of equipment for controlling reactive power and voltage are among the major challenges of power system engineering.*

The control of voltage levels is accomplished by controlling the production, absorption, and flow of reactive power at all levels in the system. The generating units provide the basic means of voltage control; the automatic voltage regulators (AVRs) control field excitation to maintain a scheduled voltage level at the terminals of the generators. Additional means are usually required to control voltage throughout the system. The devices used for this purpose may be classified as follows:

(1) Sources or sinks reactive power, such as shunt capacitors, shunt reactors, synchronous condensers, and static var compensators (SVCs).

(2) Line reactance compensators, such as series capacitors.

(3) Regulating transformers, such as tap-changing transformers and boosters.

Shunt capacitors and reactors, and series capacitors provide passive compensation. They are either permanently connected to the transmission and distribution system, or switched. They contribute to voltage control by modifying the network characteristics.

Synchronous condensers and SVCs provide active compensation; the reactive power absorbed/supplied by them is automatically adjusted so as to maintain the voltages of the buses to which they are connected.* Together with the generating units, they establish voltages at specific points in the system. Voltages at other locations in the system are determined by active and reactive power flows through various circuit elements, including the passive compensating devices.

## Part 3   Comparative Summary of Alternative Forms of Compensation

There are several methods and various devices to realize reactive power compensation. A comparative summary of alternative forms of compensation is as follows.

(1) *Switched shunt capacitor compensation* generally provides the most economical reactive power source for voltage control. It is ideally suited for compensating transmission lines if the reduction of the effective characteristic impedance, rather than the reduction of the effective line angle is the primary consideration.*

(2) *Series capacitor compensation* is self-regulating, i.e., its reactive power output increases with line loading. It is ideally suited for applications where reduction of the effective line angle is the primary consideration. It increases the effective natural load as well as the small-signal stability limit and it improves voltage regulation. It is normally used to improve system stability and to obtain the desired load division among parallel lines. Series capacitor compensation could cause sub-synchronous resonance problems requiring special solution measures. In addition, the protection of lines with series capacitors requires special attention.

(3) A *combination of series and shunt capacitors* may provide the ideal form of compensation in some cases. This allows independent control of the effective characteristic impedance and the load angle. An example of such an application is a long line requiring compensation, which causes the phase angle across the line to take a desired value so as not to adversely affect loading patterns on parallel lines.

(4) A *synchronous condenser* is a synchronous machine running without a prime mover or a mechanical load. By controlling the field excitation, it can be made to either generate or absorb reactive power. With a voltage regulator, it can automatically adjust the reactive power output to maintain constant terminal voltage. It draws a small amount of active power from the power system to supply losses.

Synchronous compensators have several advantages over static compensators. Synchronous compensators contribute to system short-circuit capacity. Their reactive power production is not affected by the system voltage. During power swings (electromechanical oscillations) there is an exchange of kinetic energy between a synchronous condenser and the power system. During such power swings, a synchronous condenser can supply a large amount of reactive power, perhaps twice its continuous rating. It has about 10% to 20% overload capability for up to 30minutes. Unlike other forms of shunt compensation, it has an internal voltage source and is better able to cope with low system voltage conditions. Some combustion turbine peaking units can be operated as synchronous condensers if required.

(5) A *static var system* (SVS) such as SVC and STATCOM is ideally suited for appli-

# Chapter 16  Voltage Regulation and Reactive Power Compensation

cations requiring direct and rapid control of voltage. It has a distinct advantage over series capacitors where compensation is required to prevent voltage sag at a bus involving multiple lines. Since shunt compensation is connected to the bus and not to particular lines, the total cost of the regulated shunt compensation may be substantially less than that for series compensation of each of the lines.

When an SVS is used to permit a high-power transfer over a long distance, the possibility of instability when the SVS is pushed to its limit must be recognized.* When operating at its capacitive limit, the SVS becomes a simple capacitor; it offers no voltage control and its reactive power drops with the square of the voltage. Systems heavily dependent on shunt compensation may experience nearly instantaneous collapse when loadings exceed the levels for which the SVS is sized. The ratings of the SVS should be based on very thorough studies which define its total mVar and the switched and dynamically controlled portions. An SVS has limited overload capability and has higher than series capacitor compensation.

## Part 4  Control by Transformers

Transformers provide an additional means of control of the flow of both real and reactive power. Our usual concept of the function of transformers in a power system is that of changing from one voltage level to another, as when a transformer converts the voltage of a generator to the transmission-line voltage. However, transformers that provide a small adjustment of voltage magnitude, usually in the range of ± 10%, and others that shift the phase angle of the line voltages are important components of a power system*. Some transformers regulate both the magnitude and phase angle.

Almost all transformers provide taps on winding to adjust the ratio of transformation by changing taps when the transformer is de-energized. A tap change can be made while the transformer is energized, and such transformers are called *load-tap-changing* (LTC) *transformers* or *under-load tap-changing* (ULTC) *transformers*. The tap changing is automatic and operated by motors that respond to relays set to hold the voltage at the prescribed level. Special circuits allow the change to be made without interrupting the current.

A type of transformer designed for small adjustments of voltage rather than for changing voltage levels is called a *regulating transformer*.

### NEW WORDS AND EXPRESSIONS

| | | | |
|---|---|---|---|
| 1. | compensation | *n.* | 补偿 |
| 2. | absorption | *n.* | 吸收 |
| 3. | excitation | *n.* | 励磁 |
| 4. | field current | | 励磁电流 |
| 5. | end-region | | 端部 |
| 6. | armature | *n.* | 电枢 |

| 7. | surge impedance | | 波阻抗，特性阻抗，浪涌阻抗 |
|---|---|---|---|
| 8. | natural load | | 自然负荷，自然功率 |
| 9. | leakage | n. | 漏电，漏磁，泄露 |
| 10. | lagging | n. | 落后，滞后，迟延 |
| 11. | prolonged | adj. | 延长的，长时期的 |
| 12. | adversely | adv | 反过来，反对地 |
| 13. | dispersed | adj. | 分散的，散布的 |
| 14. | sink | n. | 接收器 |
| 15. | booster | n. | 调压器 |
| 16. | alternative forms | | 不同形式 |
| 17. | self-regulating | | 自制的，自动调节的 |
| 18. | sub-synchronous | | 次同步的，亚同步的 |
| 19. | swing | n. | 摇摆，摆动 |
| 20. | cope with | | 应付，与…竞争 |
| 21. | voltage sag | | 电压暂降 |
| 22. | thorough | adj. | 十分的，彻底的 |
| 23. | deenergized | adj. | 失磁的，不带电的 |
| 24. | prescribed | adj. | 指定的，规定的 |

# PHRASES

**1. owing to:** 由于，因…之缘故

Example: Underground cables, <u>owing to</u> their high capacitance, have high natural loads.
地下电缆因其大电容而具有较高的自然功率。

**2. regardless of:** 不管，不顾

Example: Transformers always absorb reactive power <u>regardless of</u> their loading.
变压器一直吸收无功功率，与所带负载无关。

**3. be charged for:** 对……负责，为……付费

Example: Industrial consumers <u>are</u> normally <u>charged for</u> reactive as well as active power.
工业用户通常既要为有功付费也要为无功付费。

# COMPLICATED SENTENCES

**1. The problem of maintaining voltages within the required limits is complicated by the fact that the power system supplies power to a vast number of loads and is fed from many generating units.**

译文：将电压维持在限定范围内，这个问题的复杂性在于电力系统要给众多负载供电，同时又要从很多发电机组得到电能。或者：维持电压在所要求的限定范围内是比较复杂的，这是因为电力系统给众多负载供电，同时又从很多发电机组得到电能。

说明：这是一个被动长句，主体结构为 problem is complicated。因为主语和谓语

# Chapter 16  Voltage Regulation and Reactive Power Compensation

的修饰较多，翻译时可以处理为两个句子，用因果关系连接。为了便于理解，可以拆解为：

The problem of maintaining voltages within the required limits is complicated, because the power system supplies power to a vast number of loads and is fed from many generating units.

**2. The proper selection and coordination of equipment for controlling reactive power and voltage are among the major challenges of power system engineering.**

译文：无功和电压控制装置的合理选择与配合是电力系统工程所面临的主要挑战之一。

说明：介词 among 表示"在……之中，为……其中之一"。

**3. Synchronous condensers and SVCs provide active compensation; the reactive power absorbed/supplied by them is automatically adjusted so as to maintain the voltages of the buses to which they are connected.**

译文：同步调相机和 SVC 提供有源补偿，它们吸收/提供的无功功率都会自动调节，以维持与之相连的母线的电压。

说明：词组 so as to 表示"使得，用以"。

**4. It is ideally suited for compensating transmission lines if the reduction of the effective characteristic impedance, rather than the reduction of the effective line angle is the primary consideration.**

译文：如果首要考虑因素是减小有效特性阻抗而不是减小有效线路角度，那么它将是理想地用于输电线补偿。

说明：词组 rather than 表示"超过，胜于……"。在 if 引导的状语从句中，句型主干是 A is B。对于这种 A 部分比较长而且结构复杂而 B 部分很短的情况而言，在翻译时可以处理为 B is A。

**5. When an SVS is used to permit a high-power transfer over a long distance, the possibility of instability when the SVS is pushed to its limit must be recognized.**

译文：当 SVS 用于支持长距离大功率输电时，必须意识到将 SVS 推向其极限时失稳的可能性。

说明：句中 permit 表示"提供……机会或能力"。push to its limit 表示"推向其极限"。虽然是 static var system 的缩写形式，而且以 S 开头，但是由于 SVS 的读音以元音开头，前面的冠词用 an 而不用 a。

**6. However, transformers that provide a small adjustment of voltage magnitude, usually in the range of ±10%, and others that shift the phase angle of the line voltages are important components of a power system.**

译文：然而，变压器是电力系统重要组成部分，能够提供 ±10% 范围内的小幅度电压调整，另一些则可以改变线电压相角。

## ABBREVIATIONS (ABBR.)

1. AVR     automatic voltage regulators     自动电压调节器
2. SVC     static var compensators     静止无功补偿器
3. SVS     static var system     静止无功系统
4. STATCOM     static synchronous compensator     静止同步补偿器
5. LTC     load-tap-changing     有载调分接头
6. ULTC     under-load-tap-changing     有载调分接头

## SUMMARY OF GLOSSARY

1. 励磁状况    excitation status
   - overexcited     过励磁的
   - underexcited     欠励磁的
2. 负荷状况    load status
   - no load     空载
   - full load     满载，满负荷
   - rated load     额定负荷
3. 无功补偿    reactive power compensation
   - passive compensation     无源补偿（被动补偿）
   - active compensation     有源补偿（主动补偿）
4. 无功补偿器    reactive power compensators
   - synchronous compensator     同步补偿器
   - static compensator     静止补偿器

## EXERCISES

**1. Translate the following words or expressions into Chinese.**

(1) voltage regulation     (2) utility     (3) surge impedance
(4) field current     (5) booster     (6) sink

**2. Translate the following words or expressions into English.**

(1) 无功补偿     (2) 自动调节     (3) 自然负荷
(4) 电压暂降     (5) 同步调相机     (6) 落后，滞后

**3. Fill in the blanks with proper words or expressions.**

(1) Bus voltages will change according to _____ variation, whereas they are mainly affected by the production and absorption of _____ power at those buses.

(2) Transformers always _____ reactive power regardless of their loading; at _____ load, the shunt magnetizing reactance effects predominate; and at _____ load, the series leakage inductance effects predominate.

# Chapter 16　Voltage Regulation and Reactive Power Compensation

(3) The problem of maintaining voltages within the required limits is complicated by the fact that the power system supplies power to a vast number of _____ and is fed from many _____ units.

(4) When an SVS is used to permit a high-power transfer over a long distance, the possibility of _____ when the SVS is pushed to its limit must be recognized.

(5) A type of transformer designed for small adjustments of voltage rather than for changing voltage levels is called a _____ transformer.

## Word-Building (16)　sub-　低于；子，分，次，亚

**1. sub- [前缀] 表示：在…下；低于**

| | | |
|---|---|---|
| subway | n. | 地道，＜美＞地铁 |
| submarine | n. | 潜水艇，潜艇 |
| | adj. | 水下的，海底的 |
| subcool | vt. | 使过冷，使低温冷却 |
| submerge | v. | 浸没，淹没；潜水 |

**2. sub- [前缀] 表示：下级，分支，子，副，亚，次**

| | | |
|---|---|---|
| substation | n. | 分站，分所，变电站 |
| subroutine | n. | 子程序 |
| subsystem | n. | 子系统 |
| subclasse | n. | 子集 |
| subcode | n. | 子码 |
| subtrack | n. | 子轨道 |
| subeditor | n. | 副编辑，副主编 |
| subtitle | n. | 副题，字幕 |
| sub-synchronous | adj. | 次同步的，亚同步的 |
| sub-transient | adj. | 次暂态 |
| subsonic | adj. | 次音速的，亚音速的 |
| subdivide | v. | 再分，细分 |

# Unit 5

## Review

### I. Summary of Glossaries: Faults and System Protection

1. 回路系统　circuit systems
    - single circuit system　　　　　　　　　单回路系统
    - double circuit system　　　　　　　　　双回路系统
2. 故障分类　types of faults
    - short circuit　　　　　　　　　　　　　短路
    - open circuit　　　　　　　　　　　　　断路
    - permanent fault　　　　　　　　　　　永久故障
    - temporary fault　　　　　　　　　　　临时故障，短时故障
3. 应力与强度　stress and strength
    - electric stress　　　　　　　　　　　　静电应力，电介质应力
    - mechanical force　　　　　　　　　　机械压力
    - dielectric strength　　　　　　　　　　绝缘强度，电介质强度
    - insulation strength　　　　　　　　　　绝缘强度
4. 对称分量　symmetrical components
    - positive-sequence　　　　　　　　　　正序
    - negative-sequence　　　　　　　　　　负序
    - zero-sequencek　　　　　　　　　　　零序
5. 统计规律　statistical laws
    - random distribution　　　　　　　　　随机分布
    - normal distribution　　　　　　　　　正态分布
    - Gaussian distribution　　　　　　　　高斯分布

### II. Abbreviations (Abbr.)

1. ADP　　automated data processing　　自动数据处理
2. EMP　　electromagnetic pulses　　　　电磁脉冲
3. TOV　　temporary overvoltage　　　　临时过电压

## Special Topic (5)　Voltage Classes

The energy in the power system is carried over lines designated extra-high voltage (EHV), high voltage (HV), medium voltage (MV), and low voltage (LV). This voltage classification is made according to a scale of stand sized-voltages whose nominal values are given Table.

**Table** Voltage Classes as Applied to Industrial and Commercial Power

| Voltage class | Nominal system voltage | | | | |
|---|---|---|---|---|---|
| | Two-wire | Three-wire | | Four-wire | |
| Low voltage LV | 120 (1-p) | 120/240 ◇ (1-p) 480 600 | ◇ | 120/208 ◇ 277/480 347/600 | ◇ |
| medium voltage MV | | 2 400 4 160 4 800 6 900 13 800 23 000 34 500 46 000 69 000 | ◇ ◇ ◇ | 7 200/12 470 7 620/13 200 7 970/13 800 14 400/24 940 19 920/34 500 | ◇ ◇ ◇ ◇ |
| high voltage HV | | 115 000 138 000 161 000 230 000 | ◇ ◇ ◇ | | |
| extra-high-voltage EHV | | 345 000 500 000 735 000 -765 000 | ◇ ◇ ◇ | | |
| ultra high voltage UHV | | more | | | |

## Note:

1. All voltages are 3-phase unless indicated otherwise with a symbol 1-ph.
2. The unit of voltage is V.
3. Voltages designated by the symbol "◇" are preferred voltages.
4. Voltage class designations were approved for use by the IEEE Standards Board (September 4, 1975).

# Knowledge & Skills (5)　Promotion of Adverbial Modifier

英文科技文献中，出现在谓语动词后面的状语，在翻译时一般可将其对应内容提前到谓语动词所表示的动作之前，以使句子更通顺，更符合中文表述习惯。

**1. 简短状语：副词**

这种情况比较简单，只要在翻译时将该副词所表示的含义提到谓语动词（be 动词除外）之前即可。

例1：They are generated internally by connecting or disconnecting the system.

它们在内部由连接或断开系统产生。

说明：generated 后面除了副词 internally 之外还有一个 by 引导的介宾短语作为状语。多个状语之间的关系可以根据中文表述习惯来安排顺序。

**2. 复杂状语：状语短语**

介宾短语作状语时，在习惯上多放在句子（或对应分句）的末尾，有时为了避免限定不清，也会紧跟在谓语动词之后。为了理清句子主干，在理解时可将状语短语适当挪移，以拉近谓语动词及其宾语的距离。在翻译时再将其有关内容提到谓语动词之前。

方法一：为了理清句子主干，在理解时将状语短语抛到句子末尾。

例1：It is necessary to compute a large number of over-voltages in order to determine with some degree of confidence the statistical over-voltages on a system.

为了在一定程度上确定系统的统计过电压，有必要计算大量的过电压。

说明：with some degree of confidence 表示"带有某种程度的信心"，是状语短语。

为便于理解，有关内容可以拆解为：…determine the statistical overvoltages on a system with some degree of confidence.

在翻译时，该状语短语所表示的内容，应提前到谓语动词表示的动作之前。

例2：The distribution of breakdowns for a given gap follows with some exceptions approximately normal or Gaussian distribution, as does the distribution of over-voltages on the system.

对于给定的气隙而言，击穿的分布除了个别例外都近似遵循正态分布或高斯分布，系统中的过电压分布也是这样。

说明：with some exceptions 是状语短语，意思是"带有某些例外"。

为便于理解，有关内容可以拆解为：… follows approximately normal or Gaussian distribution with some exceptions

在翻译时，该状语短语所表示的内容，应提前到谓语动词表示的动作之前。

方法二：为了理清句子主干，在理解时将状语短语提到主语之前。符合这种情况的有 in addition to, in order to 等引导的状语短语。

例1：Above 300kV, tests include in addition to lightning impulse and the 1-minute power frequency tests, the use of switching impulse voltages.

300kV 以上，试验除了包括雷电脉冲和 1min 工频试验，还包括开关脉冲电压的使用。

说明：in addition to 表示"除了……之外"，引导一个状语短语。

为便于理解，句子可拆解为：Above 300kV, in addition to lightning impulse and the 1-minute power frequency tests, tests include the use of switching impulse voltages.

在翻译时，该状语短语所表示的内容，应提前到谓语动词表示的动作之前。对于适合作这种处理的状语而言，翻译时也可将其有关内容放在主语之前。

# Chapter 17
# Automation and Intelligence of Power System

## Part 1  Automatic Generation Control (AGC)

Automatic generation control (AGC), one of the ancillary services of the power market, can balance power consumption and generation with load dispatch among the generators.

The overall objective of the AGC is to control the electrical output of the generating units in order to supply the continuously changing customer power demand. AGC regulates the output power of electric generators within an area in response to changes in system frequency, tie-line loading, and a linear combination of these is called area control error (ACE).* In a power system, AGC <u>carries out</u> load frequency control and economic dispatch.

The key to the whole control process is the comparison between ACE and SCE. If ACE and SCE are negative and equal, then the gap in regional output is equal to the amount of expected generation exceeding actual generation, and no error signal will be generated. If both are negative and ACE is smaller than SCE, then there will be an error signal to make the $\lambda$ increase, which in turn will increase the expected power generation of the plant. Each power plant will receive a signal to increase output determined according to the economic dispatching principle.

## Part 2  Distribution Automation System (DAS)

The distribution operators' ability to act quickly and accurately in normal as well as abnormal operating conditions has a significant impact on the reliability of the supply.

***Ways to realize distribution automation***
- installing data acquisition equipment and remotely controllable devices.
- providing information processing and decision-supporting functions.

The main functions and applied technologies of the distribution automation system (DAS) are shown in Figure 17.1.

# Chapter 17    Automation and Intelligence of Power System

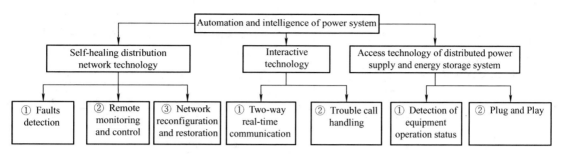

**Figure 17.1**    The Main functions and applied technologies of the distribution automation system

## Part 3    Supervisory Control and Data Acquisition (SCADA)

SCADA systems are highly distributed systems used to control geographically dispersed assets, often scattered over thousands of square kilometers, where centralized data acquisition and control are critical to system operation.

*1. Attributes*

**(1) Data acquisition**

• automatically collect information describing the system operating state by RTUs.

• pass the information to the control center in close to real-time.

**(2) Monitoring, event processing, and alarms**

• compare the measured data to normal values and limits.

• detects the change of status of switchgear and operation of protection relays.

• delivers appropriate information to the system operators through the Human-Machine Interface (HMI).

• send most critical events to the operators as alarms.

**(3) Control**

• initiate manually or automatically through the SCADA system.

• field devices control local operations.

**(4) Data storage, event log, analysis, and reporting**

The main control components of SCADA are composed of **Control Server**, SCADA Server or **Master Terminal Unit** (MTU), **Remote Terminal Unit** (RTU), **Programmable Logic Controller** (PLC), **Intelligent Electronic Devices** (IED), **Data Historian**, **Input/Output** (IO) Server.

Figure 17.2 shows the layout of the SCADA system.

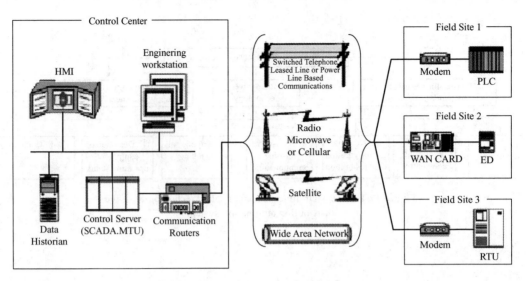

**Figure 17.2** Layout of the SCADA system

## Part 4  Energy Management System (EMS)

With the deregulation of the power industry and the development of the Smart Grid, decision-making is becoming decentralized, and coordination between different actors in various markets becomes important. The energy management system is an important part of ensuring the safe, reliable, and efficient operation of the smart grid, mainly including the management of the generation side and demand side. The generation side management includes the management of distributed power generation, energy storage system, and distribution network side. The demand-side management is mainly the management of graded load.

### 2. Function

(1) It can monitor, control, analyze, operate, make decisions, and manage the power grid, and has integrated dispatching automation, distribution network automation, substation centralized monitoring application, and analysis functions.

(2) It can provide monitoring and control functions for systems containing distributed power supply and energy storage units such as wind power generation and photovoltaic power generation. Availability, practicability, and reliability are important indicators for evaluating EMS performance.

The EMS is helpful for the conversion of power dispatching from an experience-based type to a scientific analysis-based type, in turn, may improve the security, stability, high-quality, and economic level of the power system.*

Features may vary widely from model to model, but some basic capabilities are almost universal. Several of the standard EMS capabilities are **scheduling**, **set points**, **alarms**, **safe-**

# Chapter 17 Automation and Intelligence of Power System

ties, and **basic monitoring and trending**.

## Part 5 Smart Grid

China's definition of smart grids is: based on the UHV grid as the backbone grid and the strong grid with coordinated development of power grids at all levels, using advanced communication, information, and control technology to build a unified strong smart grid characterized by informatization, automation, digitalization and interaction.

The smart grid is an advanced digital two-way power flow power system capable of self-healing, adaptive, resilient, and sustainable, with foresight for prediction under different uncertainties. Compared with traditional power grids, smart grid has the following characteristics:

(1) **Self-healing**. The smart grid monitors the power grid and reduces the probability of failure. At the same time, it can locate and automatically isolate and recover after the failure to avoid large-scale blackouts.

(2) **Reliability**. Through real-time monitoring and evaluation of the operation status of the power grid, the safety of people, equipment, and power grid can be guaranteed under different circumstances such as natural disasters, external force damage, and computer attacks, and the ability of the power grid to resist natural disasters and network attacks can be greatly improved.

(3) **Compatibility**. It can not only adapt to the centralized access of large power grids, but also support the large-scale application of distributed generation and renewable energy.

(4) **Efficiency**. Digital information technology can dynamically optimize the allocation of power resources and improve the transmission capacity and utilization of equipment; Timely adjust ments between different regions to balance the power supply gap; and support the dynamic floating electricity price system.

(5) **Interactivity**. Realizing intelligent interaction with customers, users can understand the electricity price and electricity consumption information in real-time, so as to reasonably arrange electricity consumption, and change from a single consumer to a participant in electricity trading

### NEW WORDS AND EXPRESSIONS

| 1. | ancillary | adj. | 辅助的 |
| 2. | tie-line | | 联络线 |
| 3. | economic dispatch | | 经济调度 |
| 4. | gap | n. | 缺额 |
| 5. | decision supporting | | 决策支持 |
| 6. | trouble call | | 事故呼叫, 故障呼叫 |
| 7. | scattered | adj. | 分散的 |
| 8. | switchgear | n. | 开关设备 |
| 9. | manually | adv. | 手动地, 手工地 |

| 10. | field device | n. | 现场设备 |
| 11. | historian | n. | 历史（数据）库 |
| 12. | deregulation | n. | 放松管制，解除管制 |
| 13. | decision-making | | 决策过程 |
| 14. | universal | adj. | 普遍的，通用的 |
| 15. | unified | v. | 统一，使成一体 |
| 16. | blackout | n. | 断电，停电 |

## PHRASES

**1. take into account: 考虑到，把……计算在内**

Example: This intelligent system can improve the efficiency of energy utilization while also taking into account environmental protection.

这种智能系统在提高能源利用效率的同时还兼顾环境保护。

**2. carry out: 实现，进行，执行**

Example: In a power system, AGC carries out load frequency control and economic dispatch.

在电力系统中，AGC 实现负载频率控制和经济调度。

**3. capable of: 具…能力的，能够…的**

Example: The smart grid is an advanced digital two-way power flow power system capable of self-healing, adaptive, resilient, and sustainable, with foresight for prediction under different uncertainties.

智能电网是一种先进的数字双向潮流电力系统，能够自我修复、自适应，具有弹性和可持续性，并且具有在不同不确定性下进行预测的能力。

## COMPLICATED SENTENCES

**1. AGC regulates the output power of electric generators within an area in response to changes in system frequency, tie-line loading, and a linear combination of these is called area control error (ACE).**

译文：AGC 会根据系统频率和联络线负载的变化来调节区域内发电机的输出功率，两者的线性组合称为区域控制误差（ACE）。

**2. EMS is helpful for the conversion of power dispatching from an experience-based type to a scientific analysis-based type, in turn, may improve the security, stability, high-quality, and economic level of the power system.**

译文：EMS 有助于将电网调度由经验型上升至科学分析型，进而提高电网的安全、稳定、优质和经济运行水平。

说明：词组 in turn 表示依次，这里可翻译成"进一步地"。

# Chapter 17　Automation and Intelligence of Power System

*3. Subtraction of standard or reference value from actual value to obtain the error is the accepted convention of power-system engineers and is the negative of the definition of control error found in the literature of control theory.

译文：实际值减去标准值或参考值得到误差是电力系统工程师公认的惯例，是控制理论文献中看到的误差定义的相反值。

说明：词组 subtract A from B 表示"B 减 A"。而 accepted convention 意思是"公认惯例"。

## ABBREVIATIONS (ABBR.)

| | | | |
|---|---|---|---|
| 1. | DERs | distribution energy resources | 分布式能源 |
| 2. | WAMS | wide-area measurement system | 广域测量系统 |
| 3. | AGC | automatic generation control | 自动发电控制 |
| 4. | ACE | area control error | 区域控制误差 |
| 5. | DAS | distribution automation system | 配电自动化系统 |
| 6. | SCADA | supervisory control and data acquisition | 数据采集与监控 |
| 7. | MTU | master terminal unit | 主站终端设备 |
| 8. | RTU | remote terminal unit | 远程终端设备 |
| 9. | PLC | programmable logic controller | 可编程逻辑控制器 |
| 10. | IED | intelligent electronic devices | 智能电子设备 |
| 11. | IO | input/output | 输入/输出 |
| 12. | HMI | human-machine interface | 人机接口 |
| 13. | EMS | energy management systems | 能量管理系统 |
| 14. | UHV | ultra high voltage | 特高压 |

## SUMMARY OF GLOSSARY

1. 净交换量　net interchange
   - actual net interchange　　　抗扰度，免疫性
   - schedule net interchange　　预定交换量，计划交换量
2. 差别　difference
   - error　　　误差
   - bias　　　偏差
   - offset　　　偏移量
   - deficiency　　　缺乏，不足
   - excess　　　过量，剩余
3. 在线与离线
   - on-line　　　在线
   - off-line　　　离线
4. 层次关系　layout
   - nested　　　嵌套的
   - cascading　　　级联的

# EXERCISES

**1. Translate the following words or expressions into Chinese.**

(1) economic dispatch     (2) trouble call     (3) tie-line

(4) decision supporting     (5) SCADA     (6) EMS

**2. Translate the following words or expressions into English.**

(1) 解除     (2) 数据采集     (3) 现场设备

(4) 故障检测     (5) 人机接口     (6) 配电自动化

**3. Fill in the blanks with proper words or expressions.**

(1) AGC regulates the output power of electric generators within an area in response to changes in _____ , _____ , and a linear combination of these is called _____ (ACE).

(2) The EMS is helpful for the conversion of power dispatching from an _____ type to a _____ type, in turn, may improve the security, stability, high-quality, and economic level of the power system.

## Word-Building (17)  -free, -less; -proof  无，免，抗

1. -free ［形容词后缀］，表示：无…的，免于…的

| | | | | |
|---|---|---|---|---|
| dust | n. | dust-free | adj. | 无尘的 |
| hand | n. | hands-free | adj. | 免提（电话）的 |
| loss | n. | loss-free | adj. | 无损的 |
| ice | n. | ice-free | adj. | 不冻的，无冰的 |

2. -less ［形容词后缀］，表示：无…，没有…的，不…的

| | | | | |
|---|---|---|---|---|
| brush | n. | brushless | adj. | 不带电刷的 |
| stain | n.v. | stainless | adj. | 不生锈的，无瑕疵的 |
| wire | n.v. | wireless | adj. | 无线的 |

3. -proof ［形容词后缀］，表示：防……的

| | | | | |
|---|---|---|---|---|
| dust | n. | dustproof | adj. | 防尘 |
| fire | n. | fireproof | adj. | 耐火的，防火的 |
| lightning | n. | lightning-proof | adj. | 防雷的 |
| moisture | n. | moisture-proof | adj. | 防潮的 |
| moth | n. | mothproof | adj. | 防虫的 |

# Chapter 18
# Electric Power Market

## Part 1  Overview of the Electric Power Market

The electric power market has two meanings: broad sense and narrow sense. In a broad sense, the electric power market refers to the sum of power production, transmission, use, and sales. In a narrow sense, the electric power market refers to a competitive electric power market, which is a mechanism for electricity producers, users, and buyers to trade power and its related products through negotiation, bidding, and other ways, and determine the price and quantity through market competition. The following is for the latter.

Considering different dimensions, the electric power market can be divided into several ways, such as physical power market and financial power market according to the attributes of trading products and trading intentions; According to sales behavior, it can be divided into wholesale market and retail market; According to the length of trading cycle, it can be divided into medium and long term (forward/future) market and spot market; According to the subject matter of transaction, it can be divided into energy market, ancillary services market, capacity market, transmission right market, etc.; The spot electric power market can also be subdivided into day-ahead market, intraday market, and real-time market according to the transaction time scale.

Compared with the competitive market of other energy products, the electricity market has significant characteristics. From the perspective of the short-term market, except that the demand of some large users is slightly elastic, the retail demand is rigid in the short term.

## Part 2  Physical Power Market and Financial Power Market

The power market includes the physical market and the financial market. The physical power market includes the natural resources, infrastructure, market system, and market participants related to power production, transmission, and other links, as well as the transaction, delivery, and settlement of physical goods; The financial market involves the buying and selling behavior of financial products <u>derived from</u> physical electricity, including market structure and relevant institutional arrangements, market participants, products and transactions, and also has its unique supply and demand drivers. The transactions in the power physical market mainly include forward contract transactions, spot transactions, auxiliary service transactions, etc.; The power financial market mainly conducts power derivatives trading with multi-year, annual, quarterly, monthly, and weekly trading cycles. The trading products include power futures, power options, financial transmission rights, and power price difference contracts. The electricity physical market can be divided into wholesale market and retail market according to its sales behavior. The fundamental difference between the two is that the market model and the audience are different. The audience of the wholesale market is a composite audience, which can be the terminal consumer of electric energy, or the purchase and sale of electricity, or even the producer. The buyer can conduct secondary sales; The audience in the retail market is a single audience, suitable for one-time consumption.

## Part 3  Power Market Transaction Pricing

For different trading objects, several different transaction prices are adopted in the power market, including system marginal price and node marginal price.

The system marginal price refers to the price quoted by the last power supplier (marginal unit) in spot electricity trading, which deals in electricity one by one according to the order of quotation from low to high so that the traded electricity can meet the load demand; In general, the power market requires market members to propose the quantity and price of electricity commodities to be bought/sold in a certain time in the future within a specified time, and the power trading institution determines the transaction quantity and price of both parties according to a certain optimization goal (generally speaking, maximizing social welfare). Transmission congestion is not considered in the calculation of the system's marginal price.

The settlement according to marginal pricing (MP) is the mainstream settlement method in the centralized power market. In addition, the power market can also be settled according to the actual quotation of the winning power producer (Pay as bid, PAB). Under MP, participants will automatically settle according to the most expensive bid price, while under PAB, participants have the motivation to quote according to the most expensive bid price.

When adopting a centralized or power bank trading mechanism, the system operator is also the market operator and the producers and users need to submit their bids and quotations to the system operator. The system operator is independent of all other market members and will determine the bid and quotation for winning the bid according to the principle of optimal economic efficiency, <u>taking into account</u> the security constraints of the transmission network, <u>so as to</u> realize market clearing. As part of the above process, the system operator also needs to give the market clearing price.

When congestion is taken into account, the price of electric energy will depend on the corresponding node location when the electric energy is produced or used, that is, at which

# Chapter 18　Electric Power Market

node it is injected or flowed out.* When the capacity of the interconnection line is lower than the quantity required by the free exchange, the price difference will be generated at both ends of the interconnection line because the local power generation companies need to make up the electricity required by the free exchange.* System security will form transmission constraints, which will lead to transmission network congestion. Line congestion will divide the original complete market into multiple smaller markets, resulting in location marginal price and even node marginal price.

## NEW WORDS AND EXPRESSIONS

| | | | |
|---|---|---|---|
| 1. | electric power market | n. | 电力市场 |
| 2. | physical power market | | 电力实物市场 |
| 3. | financial power market | | 电力金融市场 |
| 4. | wholesale market | | 批发市场 |
| 5. | retail market | | 零售市场 |
| 6. | future | n. | 期货，将来，前景，前途，前程 |
| 7. | spot market | | 现货市场 |
| 8. | transaction | n. | 交易，处理，业务，买卖，办理 |
| 9. | energy market | | 电能量市场 |
| 10. | ancillary services market | | 辅助服务市场 |
| 11. | capacity market | | 容量市场 |
| 12. | transmission right market | | 输电权市场 |
| 13. | day-ahead market | | 日前市场 |
| 14. | intraday market | | 日内市场 |
| 15. | real-time market | | 实时市场 |
| 16. | elastic | adj. | 有弹力的，有弹性的，橡皮圈（或带）的，灵活的，可改变的，可伸缩的 |
| 17. | derivative | n. | 派生词，衍生物，衍生字，派生物　adj.模仿他人的，缺乏独创性的 |
| 18. | system marginal price | | 系统边际电价 |
| 19. | node marginal price | | 节点边际电价 |
| 20. | quote | v. | 引用，报价，引述，举例说明，开价，出价，报价 |
| 21. | commodity | n. | 商品，有用的东西，有使用价值的事物 |
| 22. | welfare | n. | （政府给予的）福利；（个体或群体的）幸福，安全与健康 |
| | | adj. | 接受福利救济的，福利的，（从事）福利事业的 |
| 23. | transmission congestion | | 输电阻塞 |
| 24. | bid | v. | 投标，（尤指拍卖时的）出价，命令，吩咐；努力争取，企图获得，（某些牌戏中）叫牌，向（某人）问候、道别等 |
| | | n. | 出价，喊价，投标，努力，争取，叫牌 |
| 25. | market clearing | | 市场出清，结算 |
| 26. | transmission constraint | | 传输约束 |
| 27. | location marginal price | | 区位边际电价 |

## PHRASES

**1. derived from: 源自，来源于**

Example: The financial market involves the buying and selling behavior of financial products derived from physical electricity.

金融市场涉及实物电能衍生出的金融产品的买卖行为。

**2. take into account: 考虑到；so as to: 为了；以便；为使**

Example: The system operator is independent of all other market members and will determine the bid and quotation for winning the bid according to the principle of optimal economic efficiency, taking into account the security constraints of the transmission network, so as to realize market clearing.

系统运营商独立于其他所有市场成员，他会在计及输电网络安全约束的前提下，按照最优经济效率原则决定中标的投标和报价，实现市场出清。

## COMPLICATED SENTENCES

**1. When congestion is taken into account, the price of electric energy will depend on the corresponding node location when the electric energy is produced or used, that is, at which node it is injected or flowed out.**

译文：在计及阻塞的情况下，电能价格将取决于电能生产或使用时对应的节点位置，即它是在哪些节点注入或流出的。

说明：when the electric energy is produced or used 是修饰 node location 的定语从句；that is, at which node it is injected or flowed out 是对前半句的解释说明。

**2. When the capacity of the interconnection line is lower than the quantity required by the free exchange, the price difference will be generated at both ends of the interconnection line because the local power generation companies need to make up the electricity required by the free exchange.**

译文：当互联线路的容量低于自由交易所需要的数量，由于当地发电商需要弥补自由交易所需的电量，就会在互联线路两端产生价差。

说明：because the local power generation companies need to make up the electricity required by the free exchange 是前半句产生的原因，原文可写为：When the capacity of the interconnection line is lower than the quantity required by the free exchange, because the local power generation companies need to make up the electricity required by the free exchange, the price difference will be generated at both ends of the interconnection line.

## SUMMARY OF GLOSSARY

| | | |
|---|---|---|
| 1. | (electric) power market | 电力市场 |
| | physical power market | 电力实物市场 |

# Chapter 18  Electric Power Market

  financial power market        电力金融市场
  wholesale market         批发市场
  retail market          零售市场
  future market          期货市场
  spot market           现货市场
  energy market          电能量市场
  ancillary services market      辅助服务市场
  capacity market          容量市场
  transmission right market      输电权市场
  day-ahead market         日前市场
  intraday market          日内市场
  real-time market          实时市场

2. **price  classification**         **价格分类**
  system marginal price        系统边际电价
  node marginal price         节点边际电价
  location marginal price        区位边际电价

3. **terminology related to economics**    **经济学相关术语**
  wholesale            批发
  retail              零售
  future              期货
  spot               现货
  quote              报价
  bid               投标
  clear              出清，结算
  negotiation           谈判
  contract            合同
  warranty            担保；保修
  share              份额；股份
  procurement          采购
  business card          名片
  auction             拍卖

# EXERCISES

**1. Translate the following words or expressions into Chinese.**

(1) electric power market    (2) financial power market    (3) wholesale market
(4) ancillary services market   (5) capacity market      (6) intraday market
(7) system marginal price    (8) transmission congestion   (9) transmission constraint

**2. Translate the following words or expressions into English.**

(1) 零售市场        (2) 衍生物        (3) 输电阻塞
(4) 边际价格        (5) 输电权        (6) 期货市场

## 3. Fill in the blanks with proper words or expressions.

(1) System security will form _____ , which will lead to transmission network congestion. Line congestion will divide the original complete market into multiple smaller markets, resulting in _____ and even node marginal price.

(2) For different trading objects, several different transaction prices are adopted in the power market, including _____ and _____ .

# Chapter 19
# Power System Planning and Design

## Part 1  Power System Planning

Power system planning is the simulation of power system operation. In the process of planning, not only should we base on the reality and focus on the characteristics of the existing power grid, but also should consider the possible weaknesses of the future power grid and prepare for the transition. The core of the planning is to determine the construction plan, scale and timing, and the planning conclusion will deeply affect the investment decision. Its contents include **power load forecasting, power resource development, power development planning**, and **power network development planning**. The geographical wiring diagram, single line wiring diagram, and annual engineering construction project list of the power system are proposed.

Power system planning usually follows the following principles:

*1. Periodicity*

Taking the project cycle as a unit, it can accurately divide large and complex engineering projects, improve the work efficiency of power engineering, simplify the complex, and avoid excessive consumption of capital investment. Power system planning is usually divided into medium and long-term planning and short-term planning.

*2. Security*

The principle of security is the most important condition for the planning and design of the power system. If necessary, the system detection function should be equipped, such as the sensor-based power system alarm device, which can make scientific judgments for the abnormalities in the operation of the power system and give early warning for emergencies.

*3. Cost principle*

While realizing the system functions, it is also necessary to measure the cost of power system design. Scientific cost input can enable scientific planning of components, cables, and other equipment, which can not only avoid resource waste but also improve the management performance of the power system through reasonable configuration.

# Part 2　Optimization Method of Power System Planning

With the continuous expansion of the scale of the power system, the massive grid connection of renewable energy, and the continuous popularization of flexible loads such as electric vehicles, it is difficult to rely on the traditional deterministic power flow analysis method to meet the increasingly complex power system planning problems. Adopting appropriate models and algorithms for planning optimization can effectively improve the effect of power grid planning.

**1. Power grid planning**

(1) *Power grid planning model*
1) Deterministic model: transportation model, minimum cost model;
2) Reliability model: minimum cost planning, comprehensive resource planning.
(2) *Power grid planning evaluation model*: Hierarchical Structure Model.
(3) *Power grid planning solution algorithm*
1) Heuristic methods: sensitivity method, genetic algorithm, Simulated annealing(SA), list optimization, Tabu search(TS);
2) Mathematical optimization methods: linear programming, integer programming, fuzzy programming, grey theory, dynamic programming, stochastic programming, evolutionary programming, mixed integer programming, decomposition method, artificial neural network method, genetic algorithm.

**2. Power planning**

(1) *Site selection and capacity determination*
1) Bi-level coordinated planning. Aiming at minimizing the network loss, the improved particle swarm optimization algorithm of the hybrid simulated annealing algorithm is used to solve the DG location and sizing problem*;
2) The **adaptive genetic algorithm** is used to solve the DG layout planning problem;
3) The **differential evolution algorithm** and **harmonic algorithm** are combined to optimize the capacity and location of wind turbines in the distribution network on the premise of ensuring the safety and stability of wind power grid connection.
(2) *Power system load forecasting method*
neural network method, time series method, regression analysis method, support vector method, fuzzy forecasting method.

# Part 3　Content of Power System Design

**1.** *Power load forecasting and analysis*

Short-term load forecasting is mainly used to forecast the power load in the next few hours, from one day to several days. It <u>is of great significance</u> for dispatching and scheduling

# Chapter 19　Power System Planning and Design

the startup and shutdown plan, optimal unit combination, economic dispatching, optimal power flow, and power market transactions.

### 2. *Power planning*

The power supply in the power system is mainly divided into centralized power supply and local power supply. The former can dispatch power uniformly, and the latter includes small hydropower stations and corresponding generating units. The hydrological period output of different power supply locations is different, so it is necessary to conduct detailed analysis and statistics on the power supply output.

### 3. *Research on constraints*

Under the constraints of the balanced development of power and electricity, carry out the balanced calculation of power and electricity for regional projects, comprehensively analyze the results to determine the layout and scale of the power project , and set the capacity of power transformation equipment required in the power system planning*.

### 4. *Electrical calculation*

Specific electrical calculation methods include power flow calculation, stability calculation, short-circuit current calculation, and reactive compensation calculation. Taking power flow calculation as an example, power flow calculation can verify whether the system components are qualified, realize relay protection and stability calculation, and directly calculate the power loss of each node and component in the network and the actual distribution of power flow.

### 5. *Scheme comparison*

Based on the results of electrical calculation, the most reasonable power planning scheme is selected through a comprehensive analysis of safety, reliability, development adaptability, and economy. After the scheme is determined, the implementation of the power system project is started to provide more favorable data support for the engineering design discipline.

## Part 4　Wiring and Operation of Power System

The classification of main electrical wiring includes busbar and non-busbar, as shown in Figure 19.1.

### 1. *Busbar wiring*

(1) Single bus: single bus connection, single bus section connection, single bus with bypass connection;

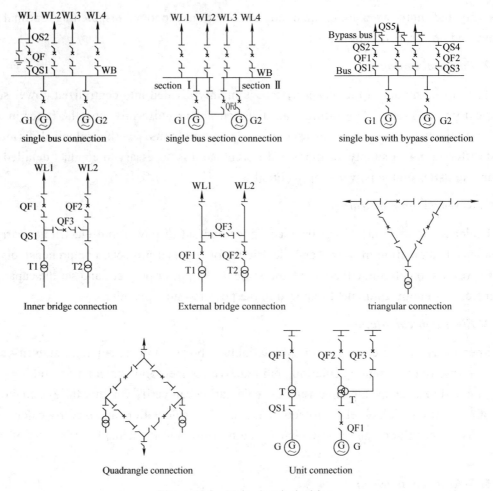

Figure 19.1  Main electrical wiring

(2) Double bus: double bus connection, double bus section connection, double bus with bypass connection, 3/2 circuit breakers connection, 4/3 circuit breakers connection, transformer bus group connection.

**2. non-busbar wiring**

unit connection, bridge connection, and angle connection.

There are four **operation states** of the power system: operation state, hot standby state, cold standby state' and maintenance state.

# Part 5  Equipment Selection and Safety Clearance

The principles to be followed when selecting electrical equipment are: select equipment according to normal working conditions and calibrate according to short circuit conditions.

When selecting the bare conductor, first determine the material of the conductor (such

# Chapter 19　Power System Planning and Design

as copper, aluminum, or aluminum alloy), and then determine the type of the conductor. The hard conductor is generally divided into rectangular, slot type, and tube type. The soft conductor is generally divided into reinforced aluminum strands, split conductors, expanded conductors, and composite conductors. According to the determined conductor model, calculate the long-term heating allowable current of the conductor, use the allowable current to select the economic current density, and conduct corona voltage verification and dynamic thermal stability verification after the type selection.

The minimum safety clearance shall be met when selecting power distribution devices. The difference between minimum safety clearance and step voltage should be distinguished here.

### 1. Minimum safety clearance

It means that the air gap will not be broken down under the maximum working voltage or internal and external overvoltage at this distance. When the maximum field strength in the slightly non-uniform electric field reaches 30kV/cm, the gap will be broken down, that is, the air will be broken down by applying 3kV voltage on the 1cm air gap.

### 2. Step voltage

When the electrical equipment is grounded, the grounding current flows to the ground through the grounding body, forming a distributed potential on the ground. At this time, if people walk around the ground short circuit point, the potential difference between their two feet (generally 0.8m) is the step voltage. When the step voltage reaches 40～50V, people will have the risk of electric shock. Once you enter the step voltage zone by mistake, you should take small steps. Do not land with both feet at the same time. It is better to jump with a one-foot walk towards the area opposite the grounding point, and gradually leave the step voltage zone.

## NEW WORDS AND EXPRESSIONS

| | | | |
|---|---|---|---|
| 1. | timing | n. | 时序，时间，定时 |
| 2. | diagram | n. | 示意图，图纸 |
| 3. | capital | n. | 资本，资金 |
| 4. | sensor | n. | 传感器 |
| 5. | abnormality | n. | 异常，反常 |
| 6. | configuration | n. | 配置，构造 |
| 7. | algorithm | n. | 算法 |
| 8. | hydrological | adj. | 水文的 |
| 9. | implementation | n. | 实施，执行 |
| 10. | calibrate | v. | 校验，校准 |
| 11. | allowable | adj. | 允许的 |
| 12. | grounding body | n. | 接地体 |
| 13. | potential | n. | 电位，电势 |

## PHRASES

**1. simplify the complex: 化繁为简**

Example: Improve the work efficiency of power engineering, simplify the complex, and avoid excessive consumption of capital investment.

提升电力工程的工作效率，化繁为简，避免消耗过多的资本。

**2. be of great significance: 具有重要意义**

Example: It is of great significance for dispatching and scheduling the startup and shutdown plan, optimal unit combination, economic dispatching, optimal power flow, and power market transactions

对于调度安排开停机计划、机组最优组合、经济调度、最优潮流、电力市场交易有着重要的意义。

**3. have the risk of: 有……的风险**

Example: When the step voltage reaches 40～50V, people will have the risk of electric shock.

当跨步电压达到40～50V时，将使人有触电危险。

## COMPLICATED SENTENCES

**1. Aiming at minimizing the network loss, the improved particle swarm optimization algorithm of the hybrid simulated annealing algorithm is used to solve the DG location and sizing problem.**

译文：以网络损耗最小为目标，混合模拟退火算法的改进粒子群算法对DG（分布式电源）选址定容问题进行求解。

说明：aim at 有"努力达到"的含义。improved 为过去分词作定语，修饰"particle swarm optimization algorithm（粒子群优化算法）"。to solve 为不定式，做主语补足语。size 作动词时有"标定……大小"之意，文中 sizing 为动名词，可翻译为定容。

**2. Under the constraints of the balanced development of power and electricity, carry out the balanced calculation of power and electricity for regional projects, comprehensively analyze the results to determine the layout and scale of the power project, and set the capacity of power transformation equipment required in the power system planning.**

译文：在电力电量平衡化发展的约束下，对区域项目进行电力、电量的平衡化计算，并对平衡结果进行综合分析，以确定电力工程的布局和规模，设定电力系统规划中所需要的变电设备容量。

说明："carry out the balanced ……"为祈使句，carry 为谓语，carry out 有"实施……，执行……"之意。determine 作动词时有"决定、测定、准确算出"之意，原文中 to determine 为不定式，做状语。required 为过去分词，作后置定语，修饰"the capacity of power transformation equipment"。

# Chapter 19  Power System Planning and Design

## ABBREVIATIONS (ABBR.)

1. AHP     analytic hierarchy process     层次分析法
2. SA     sensitivity analysis     灵敏度分析
3. DG     distributed generation     分布式电源

## SUMMARY OF GLOSSARY

1. 电气计算   electrical calculation
   - power flow calculation — 潮流计算
   - stability calculation — 稳定计算
   - short-circuit current calculation — 短路电流计算
   - reactive compensation calculation — 无功补偿计算
   - long-term heating allowable current — 长期发热允许电流
   - economic current density — 经济电流密度
   - minimum safety clearance — 最小安全净距
   - step voltage — 跨步电压
2. 金属   metal
   - copper — 铜
   - aluminum — 铝
   - aluminum alloy — 铝合金
3. 导线类型   type of the conductor
   - busbar — （有汇流）母线
   - non-busbar — 无汇流母线
   - hard conductor — 硬导体
   - rectangular — 矩形
   - slot type — 槽型
   - tube type — 管型
   - soft conductor — 软导体
   - reinforced aluminum strand — 钢筋铝绞线
   - split conductor — 分裂导线
   - expanded conductor — 扩径导线
   - composite conductor. — 组合导线
4. 算法   algorithm
   - genetic algorithm — 遗传算法
   - simulated annealing — 模拟退火
   - Tabu search — Tabu 搜索
   - fuzzy programming — 模糊规划
   - stochastic programming — 随机规划
   - Bi-level coordinated planning — 双层协调规划
   - adaptive genetic algorithm — 自适应遗传算法
   - neural network method — 神经网络法
   - regression analysis method — 回归分析法
5. 操作   operation
   - close — 闭合

| | |
|---|---|
| break-off | 断开 |
| process of switching operation | 倒闸操作 |

## EXERCISES

**1. Translate the following words or expressions into Chinese.**

(1) sensor  (2) configuration  (3) calibrate
(4) copper  (5) slot  (6) fuzzy

**2. Translate the following words or expressions into English.**

(1) 电晕  (2) 电位  (3) 补偿
(4) 短路  (5) 随机  (6) 协调

**3. Fill in the blanks with proper words or expressions.**

(1) Power system planning includes: _____.

(2) Power system planning is usually divided into _____ and _____ planning..

(3) Power flow calculation can verify whether the _____ are qualified, realize _____ and calculation, and directly calculate the _____ of each node and component in the network and the of power flow.

## Word-Building (18)　over-　过度，过分，在……之上，翻转

### 1. over-[形容词前缀]，表示：过度，过分

| confident | adj. | over-confident | adj. | 过于自信 |
|---|---|---|---|---|
| crowded | adj. | over-crowded | adj. | 过于拥挤 |
| burden | v. | over-burden | v. | 使负担过重 |
| correct | adj. | over-correct | v. | 矫枉过正 |

### 2. over-[名词前缀]，表示：在……之上

| bridge | n. | over-bridge | n. | 天桥 |
|---|---|---|---|---|
| coat | n. | over-coat | n. | 外套 |
| tone | n. | over-tone | n. | 言外之意 |

### 3. over-[动词前缀]，表示：翻转

| haul | n. | over-haul | v. | 超越对手，检修 |
|---|---|---|---|---|
| throw | v. | over-throw | v. | 打倒 |
| turn | v. | over-turn | v. | 翻转 |

# Unit 6

# Review

## I. Summary of Glossaries: Automation, Power Market and Planning

1. 净交换量　net interchange
   - actual net interchange　　　　　　　　　抗扰度，免疫性
   - schedule net interchange　　　　　　　　预定交换量，计划交换量
2. 差别　difference
   - error　　　　　　　　　　　　　　　　误差
   - bias　　　　　　　　　　　　　　　　　偏差
   - offset　　　　　　　　　　　　　　　　偏移量
   - deficiency　　　　　　　　　　　　　　缺乏，不足
   - excess　　　　　　　　　　　　　　　　过量，剩余
3. 在线与离线
   - on-line　　　　　　　　　　　　　　　在线
   - off-line　　　　　　　　　　　　　　　离线
4. 层次关系　layout
   - nested　　　　　　　　　　　　　　　嵌套的
   - cascading　　　　　　　　　　　　　　级联的
5. 电力市场　(electric) power market
   - physical power market　　　　　　　　电力实物市场
   - financial power market　　　　　　　　电力金融市场
   - energy market　　　　　　　　　　　　电能量市场
   - ancillary services market　　　　　　　辅助服务市场
   - capacity market　　　　　　　　　　　容量市场
   - transmission right market　　　　　　　输电权市场
6. 价格分类　price classification
   - system marginal price　　　　　　　　　系统边际电价
   - node marginal price　　　　　　　　　　节点边际电价
   - location marginal price　　　　　　　　区位边际电价
7. 电气计算　electrical calculation
   - power flow calculation　　　　　　　　　潮流计算
   - stability calculation　　　　　　　　　　稳定计算
   - short-circuit current calculation　　　　　短路电流计算
   - reactive compensation calculation　　　　无功补偿计算

## II. Abbreviations (Abbr.)

1. DERs    distribution energy resources    分布式能源
2. WAMS    wide-area measurement system    广域测量系统
3. AGC    automatic generation control    自动发电控制
4. ACE    area control error    区域控制误差
5. DAS    distribution automation system    配电自动化系统
6. SCADA    supervisory control and data acquisition    数据采集与监控
7. MTU    master terminal unit    主站终端设备
8. RTU    remote terminal unit    远程终端设备
9. PLC    programmable logic controller    可编程逻辑控制器
10. IED    intelligent electronic devices    智能电子设备
11. AHP    analytic hierarchy process    层次分析法
12. SA    sensitivity analysis    灵敏度分析
13. DG    distributed generation    分布式电源

# Special Topic (6)    Escape of Words

## 1. 词性不变，词义改变

① **bank**    n.    本义：银行    转义：…组
     capacitor bank    电容器组
     reactor bank    电抗器组
     transformer bank    变压器组

② **live**    adj.    本义：活的    转义：带电的
     live line    火线（相线）
     The switches enable us to isolate transformers from a live network.（带电网络）

③ **outage**    n.    本义：储运损耗    转义：断电，电力临时停供
     The protection system restricts the outage to the smallest number of customers.

④ **plant**    n.    本义：工厂，车间    转义：电厂
     thermal plant    热电厂

⑤ **utility**    n.    本义：功用，效用    转义：公用事业单位，如电力公司

⑥ **grid**    n.    本义：网格    转义：电网

⑦ **footprint**    n.    本义：足迹    转义：（设备）占地面积

⑧ **tie**    n.    本义：系带    转义：联结，联络线
     a tie between two subsystems 两个子系统之间的联络线

⑨ **reactive**    adj.    本义：反作用的    转义：无功的
     **active**    adj.    本义：积极的    转义：有功的，有源的
     **passive**    adj.    本义：被动的    转义：无源的
     reactive power 无功功率    active power 有功功率
     passive filter 无源滤波器    active filter 有源滤波器

⑩ **field**    n.    本义：场地    转义：电场，磁场；现场
     field current 励磁电流    field effect [电] 场效应
     field test 现场测量

## 2. 词性变化，词义相关。

① **monitor**　　　　n.　　本义：监控器　　v.　　转义：**监控**
　　to monitor the power quality 监控电能质量
② **charge**　　　　n.　　本义：电荷　　　v.　　转义：**充电**
　　to charge a capacitor 给电容充电，The capacitor is discharged through the resistor.

## 3. 词性变化，词义改变

① **station**　　　　n.　　本义：岗位，…站　　v.　　转义：**值班，值守**
② **power**　　　　n.　　本义：电力，功率　　v.　　转义：**供电，激励**

# Knowledge & Skills (6)　Omitting of Verbs

**1. 一个动词带有多个介宾短语时，该动词往往只在第一个位置出现，后面的可省略**

例1：The transformer insulation is subjected to high electric stress and large mechanical forces.

变压器的绝缘承受很高的静电应力和很大的机械压力。

说明：谓语表意动词 subject 带有两个并列的介宾短语，也可以看作是两个并列动词短语的省略形式，补全应为：The transformer insulation is subjected to high electric stress and is subjected to large mechanical forces.

例2：Grounding is defined as a conducting connection, whether intentional or accidental, by which an electric circuit or equipment is connected to the earth, or to some conducting body of a relatively large extent that serves in place of the earth.

接地被定义为有意或无意的导电连接，即电路或设备通过这种连接与大地或代替大地的大导体相连。

说明：谓语表意动词 subject 带有两个并列的介宾短语，即 to some conducting body 与 is connected to the earth 并列，前面省略了动词 is connected 部分。补全应为：

... an electric circuit or equipment is connected to the earth, or is connected to some conducting body of relatively large extent that serves in place of the earth.

**2. 多个并列的动词短语结构相同时，共同的谓语动词可以省略**

例1：For the purpose of coordinating the electrical stresses with electrical strengths, it is convenient to represent the overvoltage distribution in the form of the probability density function and the insulation breakdown probability by the cumulative distribution function.

为了使静电应力和电场强度配合，最好用概率密度函数的形式表示过电压，用累积分布函数表示绝缘击穿概率。

说明：it is convenient 后面有两个并列的动词不定式短语结构，由 and 连接。其动词不定式均为 to represent，只是在 the insulation breakdown probability 前省略了 to represent。补全应为：

it is convenient to represent the overvoltage distribution in the form of ... and to represent the insulation breakdown probability by the cumulative distribution function.

### 3. 其他省略情况

例 1：By definition, a PCC is the point of the public supply network, electrically nearest to a particular consumer's installation, and at which other consumer's installations are, or may be, connected.

按照定义，PCC 是公共供电网络在电气上距某一特定用户设备最近的点，其他用户的设备也连接或者可以连接到那里。

说明：at which 引导的定语从句修饰 the point。在该定语从句中，谓语部分有省略，补全应为：other consumer's installation are connected or may be connected。

# Chapter 20
# High-Voltage Direct-Current Transmission

## Part 1  Overview of High-Voltage Direct-Current Transmission

The year 1954 is generally recognized as the starting date for the modern application of High voltage direct current (HVDC) transmission when a DC line of a distance of 100km began service at 100kV from the mainland of Sweden to the island of Gotland. Since then, there has been a steady increase in the application of HVDC transmission.

Operation of a DC line began in 1977 to transmit power from a mine-mouth generating plant at Center, North Dakota to near Duluth, Minnesota, a distance of 740km. Preliminary studies showed that the DC line including terminal facilities would cost about 30% less than the comparable AC line and auxiliary equipment. This line operates at ± 250kV (500kV line to line) and transmits 500MW.

DC power can be transmitted in cables over great distances. The capacitance of a cable limits AC power transmission to a few tens of kilometers. Beyond this limit, the reactive power generated by cable capacitance exceeds the rating of the cable itself. Because capacitance does not come into play under steady-state DC conditions, there is theoretically no limit to the distance that power may be carried this way. As a result, power can be transmitted by cable under large bodies of water, where the use of AC cables is unthinkable. Direct current was chosen to transfer power under the English Channel between Great Britain and France. The use of direct current for this installation also avoided the difficulty of synchronizing the AC systems of the two countries. Furthermore, underground DC cables may be used to deliver power large urban centers. Unlike overhead lines, underground cables are invisible, free from atmospheric pollution, and solve the problem of securing rights of way.*

DC transmission has many advantages over alternating current, but DC transmission remains very limited in usage except for long lines because no DC device can provide excellent switching operations and protection of the AC circuit devices.* There is also no simple device to change the voltage level, which the transformer accomplishes for AC systems.

No network of DC lines is possible at this time because no circuit breaker is available for direct current comparable to the highly developed AC breakers. The AC breaker can extinguish the arc that is formed when the breaker opens because zero current occurs twice in each cycle. The direction and amount of power in the DC line are controlled by the converters in which grid-controlled mercury-arc devices are being displaced by the semiconductor devices.*

## Part 2  Basic DC Transmission System

A DC transmission system consists basically of a DC transmission line connecting two AC systems. A converter at one end of the line converts AC power into DC power while a similar converter at the other end reconverts the DC power into AC power. One converter acts therefore as a rectifier, the other as an inverter. More exactly, converters at the two ends of the DC lines operate both as rectifiers to change the generated alternating to direct current and as inverters for converting directly to alternating current so that power can flow in either direction.*

Stripped of everything but the bare essentials, the transmission system may be represented by the circuit of Figure 20.1. Converter 1 is a 3-phase, 6-pulse rectifier that converts the AC power of line 1 into DC power. The DC power is carried over a 2-conductor transmission line and reconverted to AC power using converter 2, acting as an inverter. Both the rectifier and inverter are line-commutated by the respective line voltages to which they are connected. Consequently, the networks can function at entirely different frequencies without affecting the power transmission between them.

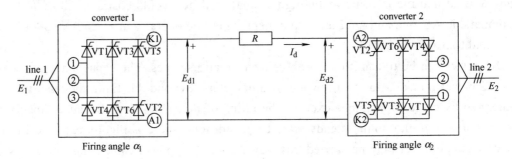

**Figure 20.1**  Elementary DC transmission system connecting 3-phase line 1 to 3-phase line 2

Power flow may be reversed by changing the firing angles $\alpha_1$ and $\alpha_2$ so that converter 1 becomes an inverter and converter 2 a rectifier. Changing the angles reverses the polarity of the conductors, but the direction of the current flow remains the same. This mode of operation is required because thyristors can only conduct current in one direction.

The DC voltages $E_{d1}$, and $E_{d2}$ at each converter station are identical, except for the *IR* drop in the line. The drop is usually so small that we can neglect it, except in so far as it affects losses, efficiency, and conductor heating.

Due to the high voltages encountered in transmission lines, each thyristor shown in Figure 20.1 is composed of several thyristors connected in series. Such a group of thyristors is often called a valve. Thus, a valve for a 50kV, 1000A converter would typically be composed

# Chapter 20  High-Voltage Direct-Current Transmission

of 50 thyristors connected in series. Each converter in Figure 20.1 would, therefore, contain 300 thyristors. The 50 thyristors in each bridge arm are triggered simultaneously, so together they act like a super-thyristor.

## Part 3  HVDC System Configurations and Components

HVDC links may be broadly classified into three categories: monopolar links, bipolar links, and homopolar links.

The basic configuration of a *monopolar HVDC link* is shown in Figure 20.2. It uses one conductor, usually of negative polarity. The return path is provided by ground or water. Cost considerations often lead to the use of such systems, particularly for cable transmission. This type of configuration may also be the first stage in the development of a bipolar system. Instead of ground return, a metallic return may be used in situations where the earth resistivity is too high or possible interference with underground / underwater metallic structures is objectionable.* The conductor forming the metallic return is at low voltage.

The *bipolar HVDC link* configuration is shown in Figure 20.3. It has two conductors, one positive and the other negative. Each terminal has two converters of equal rated voltage, connected in series on the DC side. The junctions between the converters are grounded. Normally, the currents in the two poles are equal, and there is no ground current. The two poles can operate independently. If one pole is isolated due to a fault on its conductor, the other pole can operate with the ground and thus carry half the rated load or more by using the overload capabilities of its converters and line.

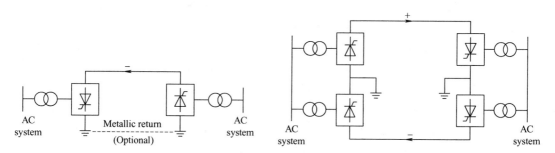

**Figure 20.2**  Monopolar HVDC link  **Figure 20.3**  Bipolar HVDC link

From the viewpoint of lightning performance, a bipolar HVDC line is considered to be effectively equivalent to a double-circuit AC transmission line. Under normal operation, it will cause considerably less harmonic interference on nearby facilities than the monopolar system. Reversal of power-flow direction is achieved by changing the polarities of the two poles through controls.

In situations where ground currents are not tolerable or when a ground electrode is not feasible for reasons such as high earth resistivity, a third conductor is used as a metallic neutral. It serves as the return path when one pole is out of service or when there is an imbalance during bipolar operation. The third conductor requires low insulation and may also serve as a shield wire for overhead lines. If it is fully insulated, it can serve as a spare.

The *homopolar HVDC link*, whose configuration is shown in Figure 20.4, has two or more conductors, all having the same polarity. Usually, a negative polarity is preferred because it causes less radio interference due to the corona. The return path for such a system is through the ground. When there is a fault on one conductor, the entire converter is available for feeding the remaining conductor(s) which, having some overload capability, can carry more than the normal power. *In contrast, a bipolar scheme is usually not feasible. The homopolar configuration offers an advantage in this regard in situations where continuous ground current is acceptable. The ground current can have side effects on gas or oil pipes, so configurations using ground return may not always be acceptable.

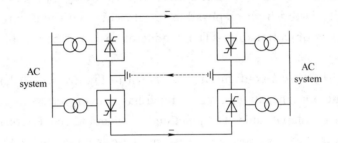

**Figure 20.4** Homopolar HVDC link

Each of the above HVDC system configurations usually has cascaded groups of several converters, each having a transformer bank and a group of valves. The converters are connected in parallel on the AC side (transformer) and in series on the DC side (valve) to give the desirable level of voltage from pole to ground.

Back-to-back HVDC systems (used for asynchronous ties) may be designed for monopolar or bipolar operation with a different number of valve groups per pole, depending on the purpose of the interconnection and the desired reliability.*

Most point-to-point (two-terminal) HVDC links involving lines are bipolar, with monopolar operation used only during contingencies. They are normally designed to provide maximum independence between poles to avoid bipolar shutdowns.

A multiterminal HVDC system is formed when the DC system is to be connected to more than two nodes on the AC network.

The main components associated with an HVDC system are shown in Figure 20.5, using a bipolar system as an example. The components for other configurations are essentially the same as those shown in the figure.

# Chapter 20  High-Voltage Direct-Current Transmission

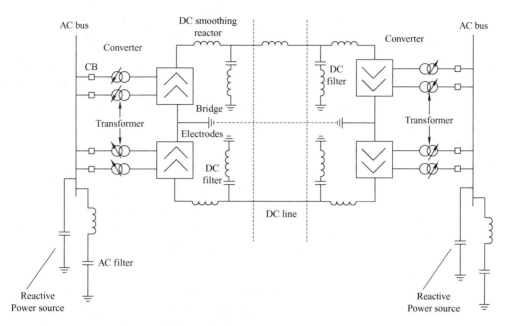

**Figure 20.5** A schematic diagram of a bipolar HVDC system identifying main components

In order to function properly, a HVDC system must have auxiliary components, in addition to the basic converters. The most important components are:
- DC smoothing reactors.
- Harmonic filters on the DC side (DC filter).
- Converter transformers.
- Reactive power source.
- Harmonic filters on the AC side (AC filter).
- Ground electrodes.
- Microwave communications link between the converter stations (not shown).
- Circuit Breakers (CB).

## Part 4  Advantages and Disadvantages of DC Transmission

The decision to use either alternating current or direct current is entirely economic in nature. Critical lengths of lines have been quoted above which the use of DC is more economical (e.g. for 750MW a critical distance of 550 ~ 750km), but these are largely dependent on the cost of the valves and associated devices which, with the increased use of direct current, will probably become cheaper.

In addition to the lower cost of DC transmission over long distances, there are other advantages. Voltage regulation is less of a problem since at zero frequency $\omega L$ is no longer a factor, whereas it is the chief contributor to voltage drop in an AC line.* Another advantage of direct current is the possibility of monopolar operation in an emergency when one side of a bipolar line becomes grounded.

DC power can be controlled much more quickly. For example, power in the megawatt range can be reversed in a DC line in less than one second. This feature makes it useful to operate DC transmission lines in parallel with existing AC networks. When instability is about to occur (due to a disturbance on the AC system), the DC power can be changed in amplitude to counteract and dampen out the power oscillations. Quick power control also means that DC short circuit currents can be limited to much lower values than those encountered on AC networks.

Still, another advantage of direct current is the smaller amount of right of way required. The distance between the two conductors of the North Dakota-Duluth 500kV line is 25 feet. The 500kV AC line has 60.5 feet between the outside conductors. Another consideration is the peak voltage of the AC line which is $\sqrt{2} \times 500 = 707\text{kV}$. So the line requires more insulation between the tower and conductors as well as greater clearance above the earth.

In two cases there are strong technical reasons for the use of DC transmission.

(a) For the connection of large systems through links of small capacity. An example is the Britain-France Cross-Channel link where slightly different frequencies in the two large systems would produce serious problems of power transfer control in the small capacity link. A DC line is an asynchronous or flexible link between two rigid systems.

(b) Where high-voltage underground cables are needed for reasonable transmission distances. The limitations of cables due to charging current with AC transmission have been discussed and to increase lengths either artificial reactors or DC must be used. DC transmission by cables inland may be expected to take place in areas where amenity consideration restricts the use of overhead lines. The use of cables for cross-channel crossings is well established. A comparison between the use of AC and DC with underground cables is interesting. Six 275kV, 3 in$^2$, cables in two groups of three in horizontal formation (total width of trench 5.2m) in soil have an AC capacity of 1520MVA. Two cables at ±500kV DC have a capacity of 1600MW with a trench width of only 0.68m.

Further advantages in the use of DC are as follows: the corona loss in a DC line operating at a voltage corresponding to the peak value of the equivalent alternating voltage is substantially less than for the AC line. This is important, not so much because of the loss of power, but due to the resulting interference with radio and television.* Practices have shown that the fault levels in an AC system with a DC link operating are less than with the link replaced by an AC equivalent. This is of great importance when the use of higher voltages and many interconnections have greatly increased the fault MVA to be withstood by circuit breakers.

The disadvantages of DC transmission are as follows:

1) The much more onerous conditions for circuit breaking when the current does not reduce to zero twice a cycle. Because of this, switching is not carried out on the DC link but is effected using the terminal rectifiers and inverters. This severely hampers the creation of an interconnected DC system with tee junctions. However, the use of multiple links with con-

# Chapter 20  High-Voltage Direct-Current Transmission

verter stations in series or parallel with each other and with the valves used as switches is likely.

2) Voltage transformation has to be provided on the AC sides of the system.

3) Rectifiers and inverters absorb reactive power which must be supplied locally.

4) DC converting stations are much more expensive than conventional AC substations.

## NEW WORDS AND EXPRESSIONS

| 1.   | mine-mouth       | n.   | 矿山口 |
| --- | --- | --- | --- |
| 2.   | preliminary      | adj. | 预备的，初步的 |
| 3.   | auxiliary        | n.   | 辅助设备，附属设备 |
| 4.   | unthinkable      | adj. | 不能想象的，不可思议的 |
| 5.   | rights of way    |      | 公用事业用地，公用道路 |
| 6.   | mercury-arc      |      | 汞弧 |
| 7.   | semiconductor    | n.   | 半导体 |
| 8.   | rectifier        | n.   | 整流器 |
| 9.   | inverter         | n.   | 逆变器 |
| 10.  | strip of         |      | 剥夺 |
| 11.  | line-commutated  |      | 线换向的 |
| 12.  | respective       | adj. | 分别的，各自的 |
| 13.  | thyristor        | n.   | 半导体闸流管，晶闸管 |
| 14.  | valve            | n.   | 阀 |
| 15.  | bridge arm       | n.   | 桥臂 |
| 16.  | trigger          | vt.  | 触发 |
| 17.  | simultaneously   | adv  | 同时地 |
| 18.  | link             | n.   | 线路，连接 |
| 19.  | monopolar        | adj. | 单极的 |
| 20.  | bipolar          | adj. | 双极的 |
| 21.  | homopolar        | adj. | 同极的 |
| 22.  | junction         | n.   | 接头，连接处 |
| 23.. | double-circuit   |      | 双回路 |
| 24.  | reversal         | n.   | 颠倒，反转，反向 |
| 25.  | feasible         | adj. | 可行的，切实可行的 |
| 26.  | spare            | n.   | 备用品 |
| 27.  | side effect      |      | 副作用 |
| 28.  | cascaded         | adj. | 级联的 |
| 29.  | transformer bank |      | 变压器组 |
| 30.  | smoothing        | n.   | 滤波 |
| 31.  | filter           | n.   | 滤波器 |
| 32.  | counteract       | vt.  | 抵消，中和 |
| 33.  | dampen           | vt.  | 消除，抑制 |

| 34. | rigid | adj. | 坚挺的，刚性的 |
| 35. | artificial | adj. | 人为的，人工的 |
| 36. | amenity | n. | 舒适 |
| 37. | trench | n. | 管沟，电缆沟 |
| 38. | onerous | adj. | 有麻烦的或有压力的 |
| 39. | hamper | v. | 妨碍，牵制 |
| 40. | tee junction | | T 型结点 |

## PHRASES

**1. come into play: 开始活动，发生影响**

Example: Because capacitance does not come into play under steady-state DC conditions, there is theoretically no limit to the distance that power may be carried this way.

因为在稳态直流条件下电容不会产生影响，理论上用这种方式输送功率没有距离限制。

**2. strip of: 剥夺**

Example: Stripped of everything but the bare essentials, the transmission system may be represented by the circuit of Figure 20.1.

去除不带多余部分的基本要素之外的任何内容，输电系统可用图 20.1 所示的电路来表示。

**3. insofar as: 到这个程度 [范围]；在…情况下；在…范围内；只要…**

Example: The drop is usually so small that we can neglect it, except in so far as it affects losses, efficiency, and conductor heating.

该压降通常很小从而可以忽略，除非大到影响损耗、效率和导体发热。

**4. in nature: 实际上，本质上**

Example: The decision to use either alternating current or direct current is entirely economic in nature.

决定用交流还是直流本质上完全是经济上的考虑。

## COMPLICATED SENTENCES

**1. Unlike overhead lines, underground cables are invisible, free from atmospheric pollution, and solve the problem of securing rights of way.**

译文：和架空线不同，地下电缆是看不见的，免受大气污染，并解决了公用通道的安全问题。

说明：unlike 意思是"不像，与……不同"。free from 表示"没有……的，免于"。rights of way 指的是公用事业用地，例如铁路线、公路线、管道线建设所需的土地。

# Chapter 20　High-Voltage Direct-Current Transmission

**2. DC transmission remains very limited in usage except for long lines because no DC device can provide excellent switching operations and protection of the AC circuit devices.**

译文：直流输电除用于长线（输电）以外在应用上仍然十分有限，这是因为没有直流设备能够提供交流装置所具有的卓越的开关操作和保护功能。

说明：词组 except for 表示"除……以外"。

**3. The direction and amount of power in the DC line are controlled by the converters in which grid-controlled mercury-arc devices are being displaced by the semiconductor devices.**

译文：直流线路上功率的流向和数量用换流器控制，其中栅控汞弧设备在逐渐被半导体整流器取代。

说明：are being displaced 是现代进行时的被动语态，表示"正在被取代"。

**4. More exactly, converters at the two ends of the DC lines operate both as rectifiers to change the generated alternating to directly current and as inverters for converting direct to alternating current so that power can flow in either direction.**

译文：更为准确地说，直流线路两端的换流器都既可作为整流器将产生的交流变为直流，也可作为逆变器将直流转换为交流，从而功率可以向每个方向流动。

说明：both as...and as 表示"既可作为……又可作为"。句式 so that 表示"从而，因而"。

**5. Instead of ground return, a metallic return may be used in situations where the earth resistivity is too high or possible interference with underground / underwater metallic structures is objectionable.**

译文：在地电阻率太大或者对地下/水下金属结构可能造成干扰的场合，可以用金属回路取代地回路。

说明：词组 interfere with 意思是"干扰，妨碍…"，在这里其名词形式 interference with 表示"对…的干扰"。词组 instead of 表示"代替"。objectionable 意思是"引起反对的，要不得的"。

**6. When there is a fault on one conductor, the entire converter is available for feeding the remaining conductor(s) which, having some overload capability, can carry more than the normal power.**

译文：当一根导线出现故障时，整个换流器可用于给剩余导线供电、剩余导线具有一定的过载能力，可承载比正常功率更多的功率。

说明：句中 is available for 意思是"可以用来"。having some overload capability 与 can 后面的内容具有相同的含义，只是表现形式不一样。

7. Back-to-back HVDC systems (used for asynchronous ties) may be designed for monopolar or bipolar operation with a different number of valve groups per pole, depending on the purpose of the interconnection and the desired reliability.

译文：背靠背 HVDC 系统（用于不同步的联络线）可设计为单极或双极运行，每极有不同数目的阀组，这取决于互联的目的和期望的可靠性。

说明：monopolar or bipolar operation 中，monopolar 的后面省略了 operation。词组 designed for 表示"设计用于"。

8. Voltage regulation is less of a problem since at zero frequency $\omega L$ is no longer a factor, whereas it is the chief contributor to voltage drop in an AC line.

译文：既然频率为零时 $\omega L$ 不再起作用，电压调节也就不成问题了，而在交流线上 $\omega L$ 却是电压下降的主要原因。

说明：less of a problem 可以理解为"不成问题"，no longer 意思是"不再"。

9. This is important, not so much because of the loss of power, but due to the resulting interference with radio and television.

译文：这很重要，不只因为功率损耗，而是由于其带来的无线电和电视干扰。

说明：so much 在此表示"就只那么多"。

## ABBREVIATIONS (ABBR.)

1.  HVDC          high voltage direct current          高压直流（输电）
2.  CB            circuit breaker                      断路器

## SUMMARY OF GLOSSARY

1. 换流器　converters
   - converter　　　　　　　　　　　　　　换流器，变流器
   - rectifier　　　　　　　　　　　　　　整流器
   - inverter　　　　　　　　　　　　　　逆变器
2. HVDC 链的分类　categories of HVDC links
   - monopolar link　　　　　　　　　　　单极型
   - bipolar link　　　　　　　　　　　　双极型
   - homopolar link　　　　　　　　　　　同极型
3. HVDC 连接　HVDC connections
   - back-to-back　　　　　　　　　　　　背靠背的
   - point-to-point　　　　　　　　　　　点对点的
   - multiterminal　　　　　　　　　　　多端的
4. 连接方式　connection manner
   - in parallel　　　　　　　　　　　　并联的［地］，并行的［地］
   - in series　　　　　　　　　　　　　串联的［地］，串行的［地］

# Chapter 20  High-Voltage Direct-Current Transmission

## EXERCISES

**1. Translate the following words or expressions into Chinese.**

(1) installation  (2) converter  (3) auxiliary equipment
(4) rights of way  (5) valve group  (6) side effect
(7) smoothing reactor  (8) filter  (9) clearance to ground

**2. Translate the following words or expressions into English.**

(1) 晶闸管  (2) 半导体  (3) 逆变器
(4) 桥臂  (5) 双回路  (6) T 型结点

**3. Fill in the blanks with proper words or expressions.**

(1) _____ at the two ends of the DC lines operate both as _____ to change the generated alternating to direct current and as _____ for converting directly to alternating current so that power can flow in either direction.

(2) HVDC links may be broadly classified into three categories: _____ links, _____ links and _____ links.

(3) Each of the above HVDC system configurations usually has _____ groups of several converters, each having a transformer bank and a group of valves. The converters are connected in _____ on the AC side (transformer) and in _____ on the DC side (valve) to give the desirable level of voltage from pole to ground.

# Word-Building (19)   semi-; super-   子，分，次，亚

## 1. semi-   [前缀] 表示：半；部分；不完全

| | | |
|---|---|---|
| semiautomatic | *adj.* | 半自动的 |
| semicircle | *n.* | 半圆形 |
| semiconductor | *n.* | 半导体 |
| semidiameter | *n.* | 半径 |
| semiproduct | *n.* | 半成品 |

## 2. super-   [前缀] 表示：超；高级；在…之上

| | | |
|---|---|---|
| superconductor | *n.* | 超导（电）体 |
| supercountry | *n.* | 超级大国 |
| supercritical | *adj.* | 超临界的 |
| superenergy | *n.* | 超高能量 |
| superhigh | *adj.* | 超高的 |
| superhuman | *n.* | 超人 |
| superimpose | *vt.* | 添加，重叠，叠合 |
| supersonic | *n. (adj.)* | 超音速（的） |
| super thyristor | *n.* | 超级晶闸管 |

# Chapter 21
# Flexible AC Transmission System

## Part 1  Concept and Development of FACTS

Today's power transmission and distribution systems face increasing demands for more power, better quality with higher reliability and at lower cost as well as low environmental impact.* The application of power electronics in new configurations known as FACTS (Flexible AC Transmission System) offers the possibility of meeting such demands.

FACTS technology uses advanced power semiconductor switching techniques to provide dynamic voltage support, power system stabilization, and enhanced power quality for transmission and distribution system applications.

FACTS devices are routinely employed in order to enhance the power transfer capability of the otherwise underutilized parts of the interconnected network.

The use of power electronic-based apparatus at various voltage levels in electric energy systems is becoming increasingly widespread due to rapid progress in power electronic technology.

The most often used power electronic elements include
- GTO, Gate Turn-Off thyristor;
- IGBT, Insulated Gate Bipolar Transistor;
- IGCT, Integrated Gate Commutation Transistor.

## Part 2  SVC Family

A Static Var Compensator (SVC), according to the basic terminology established by CIGRE, is a shunt-connected static var generator (SVG) whose output is varied so as to maintain or control specific parameters of the electric power system. The term "static" is used to indicate that SVCs, unlike synchronous compensators, have no moving or rotating main components. Thus an SVC consists of static var generator or absorber devices and a suitable control device. Presently, in practical compensators, the generation and control of the reactive

# Chapter 21    Flexible AC Transmission System

power output of a static var generator is accomplished exclusively by thyristor valves in conjunction with capacitor and /or reactor banks.

The capacitor banks are either a fixed amount or are varied in steps by thyristor switching. The reactors are varied through thyristor switching. Based on these principles, various configurations of SVCs have been developed and applied. They are characterized by a fast response, high reliability, low operating cost, and flexibility. The following types of SVCs can be identified:

- Thyristor-Controlled Reactor (TCR);
- Thyristor-Switched Reactor (TSR);
- Thyristor-Switched Capacitor (TSC);
- Thyristor-Controlled Transformer (TCT);
- Self-or Line-Commutated Transformer (SCT/LCT);
- Self-or Line-Commutated Converter (SCC/LCC);
- Any Combination of the above.

The TCR compensator consists of a combination of thyristor-controlled reactors with fixed shunt capacitor bank(s). The reactors themselves also have fixed impedances but the fundamental frequency component of the current through them is controlled by thyristor valves, giving variable impedance. The branch current is controlled by phase angle control of the firing pulses to the thyristors, that is, the voltage across the reactors is the full system voltage at 90 degrees firing angle and zero at 180 degrees.
* The current through the reactors is the integral of the voltage, (Figure 21.1), thus it is fully controllable with the thyristor valves between the natural value given by the reactor impedance and zero.

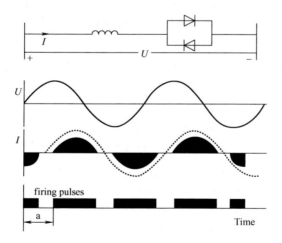

**Figure 21.1**  TCR fundamental principles

The current in the reactor can be controlled to vary from maximum to zero by controlling the firing delay angle to the thyristor valves. Hence the var output of a TCR compensator can vary continuously. TCR generates harmonics because the firing angle control results in a non-sinusoidal current waveform in the reactor. It is often necessary to remove the harmonics generated by the TCR with filters. Usually, the fixed capacitor bank(s) will be tuned to work as filters.

The TSC compensator consists of one or more thyristor-switched capacitor branches. The capacitor switching (connecting) takes place at that specific instant in a cycle at which the conditions for minimum transients are satisfied, that is when the voltage across the thyristor valve is zero or minimum. * The firing delay angle control is not applicable to capacitors. A TSC branch can only provide a step change in its reactive current (maximum or zero) and as a result, it does not generate harmonics. Therefore, a TSC compensator is a variable reactive power source with stepwise control.

Unlike mechanical switching devices such as breakers, the time between switching for the thyristor switchers in an SVC are limited only by the response time of the compensator. The response time of SVCs is in the range of 2-3 cycles which makes SVCs well suited for applications in which fast and repetitive control of reactive power is required. *

In addition to the shun-connected SVCs, there are also some similar devices in series-connection such as TCSC (Thyristor Controlled Series Compensator).

The above discussion is intended as a summary description of static var compensators. The reader is referred to some special documents for a more detailed theoretical treatment of the principals involved.

# Part 3　The Voltage Source Converter

The voltage source converters (VSC), especially voltage source inverters (VSI), are used in many FACTS devices. In order to illustrate how the VSI behaves we refer to the basic bridge connection that forms the converter, as shown in Figure 21.2.

The idea then is to create sinusoidal-like voltages at the three output terminals, from the assumed constant DC voltage across the capacitors, such that the current drawn by the converter circuit meets the identified objectives. The controllable elements in the circuit in Figure 21.2, for instance, IGBTs, must alternately connect the phase output terminals to the respective DC terminal, or the midpoint between the capacitors. In doing so they will produce a square-wave type of waveform, as each IGBT constitutes a switch that can take two states, either conducting (as a short-circuit) or blocking (open circuit). It shall be noted that this voltage shall be generated independently of the phase relation of the current that will flow to the converter bridge. The diodes that are connected antiparallel to each IGBT will ensure that there is always a path for the current to flow.

To show the function of the three-level converter, a simplified scheme is shown in Figure 21.3, where all the valves have been changed into bidirectional switches that connect the phase outputs to one out of the three potentials on the DC side. For this converter, the phase connections a, b, and c can have the same potential as either the positive or the negative terminal of the DC side or its midpoint, i.e., three possible values. Hence the name: three-level

# Chapter 21 Flexible AC Transmission System

converter.

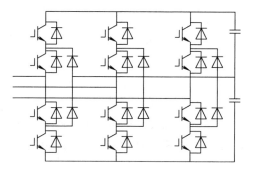

Figure 21.2 Basic 3-level Voltage Source Circuit

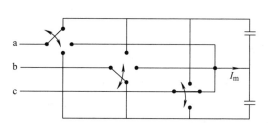

Figure 21.3 Simplified model of 3-level converter

One of the fundamental properties of the VSC is its ability to control current by applying a voltage across a reactor. The VSC obtains the voltage from the DC side capacitor. An important aspect is, however, that the capacitor does not primarily work as energy storage. Instead, under balanced conditions, all switchings of semiconductors lead to currents being circulated within the three phases. Under unbalanced conditions, the DC side capacitor will be loaded with some 2nd harmonic currents.

## Part 4 STATCOM

The Static Synchronous Compensator (STATCOM) is a principal state-of-the-art FACTS equipment and is now a commercially available additional tool for use by system planners and designers for shunt reactive power compensation at either the transmission level for voltage regulation and system stability improvement or at the distribution level for voltage regulation, power factor correction and/or improvement of power quality.*

The STATCOM is a solid-state DC to AC switching power converter that consists of a VSC connected in shunt with the system. Figure 21.4 shows the simplified equivalent circuit of the STATCOM, including a DC-side capacitor, an inverter based on power electronics switches, and series resistance and inductance. The inverter is connected to capacitor C which supplies the DC voltage. The resistance $R$ accounts for the sum of the transformer winding resistance loss and the inverter conduction losses. The inductance $L$ accounts for the leakage inductance of the transformer.

When the fundamental positive component is used as the synchronizing signal, the output voltage of the inverter is expressed as

$$\begin{bmatrix} u_{ia} \\ u_{ib} \\ u_{ic} \end{bmatrix} = \sqrt{\frac{2}{3}} \cdot kU_{dc} \cdot \begin{bmatrix} \cos(\omega_0 t + \delta) \\ \cos\left(\omega_0 t + \delta - \frac{2}{3}\pi\right) \\ \cos\left(\omega_0 t + \delta + \frac{2}{3}\pi\right) \end{bmatrix}$$

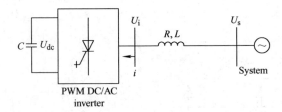

**Figure 21.4** Equivalent one-line circuit diagram of STATCOM

where $V_{dc}$ is the inverter DC-side voltage, $k$ is the ratio of the inverter output line-line voltage to the DC-side voltage, and $\delta$ is the phase angle by which the STATCOM equipment leads the system voltage.

The AC-side circuit equations can be expressed in terms of the instantaneous variables:

$$\frac{d}{dt}\begin{bmatrix} i_a \\ i_b \\ i_c \end{bmatrix} = \begin{bmatrix} -\frac{R}{L} & 0 & 0 \\ 0 & -\frac{R}{L} & 0 \\ 0 & 0 & -\frac{R}{L} \end{bmatrix}\begin{bmatrix} i_a \\ i_b \\ i_c \end{bmatrix} + \frac{1}{L}\begin{bmatrix} u_{sa} - u_{ia} \\ u_{sb} - u_{ib} \\ u_{sc} - u_{ic} \end{bmatrix}$$

Independent of the application, proper operation of any STATCOM also requires that the voltage level on its DC terminals be regulated.

The compensator supplies reactive power to the system when it is operated in the capacitive mode. In contrast, when it operates in the inductive mode, the STATCOM absorbs reactive power from the systems. But either operated in inductive or capacitive mode, the phase angle between the source and compensator will be other than 90 degrees. This is because the compensator needs to absorb active power from the source to supply the losses and support the DC link voltage.

The STATCOM can be treated as a synchronous machine with controllable voltage. The voltage can be controlled both in amplitude, phase and frequency, with full independence between the three attributes. In addition, the VSC modulated with high-frequency pulse width modulation (PWM), is capable of synthesizing also a negative sequence voltage.

As with all static FACTS devices, the STATCOM has the potential to be exceptionally reliable but with the added capability to:*

- Sustain reactive current at low voltage (constant current not constant impedance);
- Reduce land use and increase relocatability (footprint: 40% of SVC);
- Be developed as a voltage and frequency support (by replacing capacitors with batteries as energy storage).

Although currently being applied to regulated transmission voltage to allow greater power flow in a voltage-limited transmission network in the same manner as an SVC, the STATCOM has further potential. By giving an inherently faster response and greater output to a system with a depressed voltage, the STATCOM offers improved quality of supply.

As for applications in distribution systems, where they are called D-STATCOM, sever-

# Chapter 21　Flexible AC Transmission System

al control objectives are of interest and can to a high degree be fulfilled simultaneously. The identified objectives are:

- Compensation for unbalanced loads;
- Power factor correction;
- Compensation of voltage variety;
- Elimination of harmonics.

## Part 5　Other Devices Based on VSC

A *dynamic voltage restorer* (DVR) is a series device that generates an AC voltage and injects it in series with the supply voltage through an injection transformer to compensate for the voltage sag. The injected voltage and load current determine the power injection of the DVR. The schematic diagram of a typical DVR is shown in Figure 21.5. The circuit on the left-hand side of the DVR represents the Thevenin equivalent circuit of the system. The system impedance $Z_{th} = R_{th} + jX_{th}$ depends on the fault level of the load bus. When the system voltage ($V_{th}$) drops, the DVR injects a series voltage $V_{DVR}$ through the injection transformer so that the desired load voltage magnitude $V_L$ can be maintained.

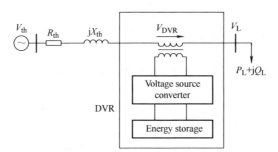

**Figure 21.5**　Schematic diagram of a DVR

The series-injected voltage of the DVR can be written as

$$V_{DVR} = V_L + Z_{th}I_L - V_{th} \qquad (1)$$

Here $I_L$ is the load current and is given by

$$I_L = \left(\frac{P_L + jQ_L}{V_L}\right)^* \qquad (2)$$

When $V_L$ is considered as a reference, Equ. (1) can be rewritten as

$$V_{DVR}\angle\alpha = V_L\angle 0 + Z_{th}I_L\angle(\beta - \theta) - V_{th}\angle\delta \qquad (3)$$

Here $\alpha, \beta$, and $\delta$ are the angles of $V_{DVR}, Z_{th}$ and $V_{th}$, respectively, and the load power factor angle $\theta = \arctan(Q_L / P_L)$.

The complex power injection of the DVR can be written as

$$S_{DVR} = V_{DVR}I_L^* \qquad (4)$$

It may be mentioned here that when the injected voltage $V_{DVR}$ is kept in quadrature with $I_L$, no active power injection by the DVR is required to correct the voltage. It requires the injection of only reactive power and the DVR itself is capable of generating the reactive power. Note that $V_{DVR}$ can be kept in quadrature with $I_L$ only up to a certain value of voltage sag and beyond which the quadrature relationship cannot be maintained to correct the voltage sag. For such a case, injection of active power into the system is essential. The injected active power must be provided by the energy storage system of the DVR. On the other hand, when the magnitude of the injected voltage is minimized, the desired voltage correction can be achieved with minimum apparent power injection into the system. This aspect of voltage correction is also very important because it minimizes the size of the injection transformer.

*Active power filters* (APF) are designed for the compensation of harmonics caused by nonlinear loads in the power system. The circuit configurations of APFs are similar to that of STATCOMs, which consist of an inverter based on power electronic switching elements and converts the DC voltage on a capacitor to AC voltage on the AC side.

As the new generation devices for harmonic suppression and reactive power compensation devices, their control system are complicated, including signal detection, stable control, and bottom driver control.

Since the major object of APFs is to provide compensation currents for the harmonic components of system currents, the required modulation frequencies used for PWM in APFs are usually much higher than those in STATCOMs.

## NEW WORDS AND EXPRESSIONS

| | | | |
|---|---|---|---|
| 1. | power electronics | | 电力电子 |
| 2. | flexible | adj. | 灵活的，柔性的 |
| 3. | enhance | vt. | 增强，提高 |
| 4. | underutilized | adj. | 未充分利用的 |
| 5. | commutation | n. | 换向 |
| 6. | terminology | n. | 术语，术语学 |
| 7. | exclusively | adv | 排外地，专有地 |
| 8. | firing pulse | | 点火脉冲 |
| 9. | nonsinusoidal | adj. | 非正弦的 |
| 10. | tune | vt. | 调谐 |
| 11. | stepwise | adj. | 逐步（的），逐渐（的），分步（的） |
| 12. | repetitive | adj. | 重复的，反复性的 |
| 13. | square-wave | | 方波 |
| 14. | constitute | vt. | 构成，组成 |
| 15. | diode | n. | 二极管 |
| 16. | antiparallel | adj. | 反平行的，反并联的 |
| 17. | energy storage | | 储能 |
| 18. | state-of-the-art | | 达到最新技术发展水平的 |
| 19. | solid-state | | 固态的，使用电晶体的 |

# Chapter 21  Flexible AC Transmission System

| 20. | conduction loss | | 传导损耗，导通损耗 |
|---|---|---|---|
| 21. | lead | vt. | 领先，超前 |
| 22. | sustain | vt. | 支撑，维持 |
| 23. | relocatability | n. | （可）再定位，浮动 |
| 24. | footprint | n. | 覆盖区，脚印 |
| 25. | batteries | n. | 电池 |
| 26. | depressed | adj. | 降低的 |
| 27. | variety | n. | 变动 |
| 28. | sag | n. | 暂降，凹陷 |
| 29. | Thevenin | | 戴维南等效 |
| 30. | quadrature | n. | 正交 |
| 31. | suppression | n. | 镇压，抑制 |

## PHRASES

**1. so as to: 使得，以致**

Example: A Static Var Compensator, according to the basic terminology established by CIGRE, is a shunt-connected static var generator whose output is varied <u>so as to</u> maintain or control specific parameters of the electric power system.

按照 CIGRE 制定的基本术语，静止无功发生器是并联的静止无功发生器，可改变其输出以至维持或控制电力系统中的特定参数。

**2. in conjunction with：与…协力；联合**

Example: Presently, in practical compensators, the generation and control of the reactive power output of a static var generator is accomplished exclusively by thyristor valves <u>in conjunction with</u> capacitor and /or reactor banks.

近来，在实际的补偿器中，静止无功发生器的无功输出的产生与控制，专门由晶闸管阀联合电容器或电抗器组实现。

**3. in steps: 分级，分步；(注意与表示"同步"的 in step 区别)**

Example: The capacitor banks are either a fixed amount or are varied <u>in steps</u> by thyristor switching.

电容器组要么是固定数量，要么是通过晶闸管投切可分级变化。

**4. be (not) applicable to: (不)适用于**

Example: The firing delay angle control <u>is not applicable to</u> capacitors.
点火延迟角控制不适用于电容。

**5. other than: 不同于，除了**

Example: But either operated in inductive or capacitive mode, the phase angle between the source and compensator will be <u>other than</u> 90 degrees.

但是不论运行在感性或容性模式，电源和补偿器之间的相角都不会是 90 度。

**6. as for：至于，就……方面来说　to a high degree：非常好地，高度地**

Example: As for applications in distribution systems, where they are called D-STATCOM, several control objectives are of interest and can to a high degree be fulfilled simultaneously.

至于在配电系统中的应用，在那里它们被称为 D-STATCOM，几个控制目标比较重要并且能很好地同时实现。

## COMPLICATED SENTENCES

**1. Today's power transmission and distribution systems face increasing demands for more power, better quality with higher reliability and at lower cost as well as low environmental impact.**

译文：如今的输配电系统面临日益增长的需求，要求功率更大，可靠性更高，质量更好，既要对环境影响小又要成本更低。

说明：power transmission and distribution systems 意思是"电力输送和分配系统"，按照我国电力行业的习惯说法可以简称为"输配电系统"。as well as 表示"在…的同时也…"。介词 for 带有多个并列的宾语，这些宾语都有各自的定语，这样的名词短语在翻译时可以处理为主谓结构，译文更容易符合中文习惯。

**2. The branch current is controlled by phase angle control of the firing pulses to the thyristors, that is, the voltage across the reactors is the full system voltage at 90 degrees firing angle and zero at 180 degrees.**

译文：通过对晶闸管点火脉冲的相角控制来控制支路电流，即，电抗器上的压降在点火角为 90 度时为全部系统电压，在 180 度时为零。

说明：that is 表示"即，换句话说，就是说"。that is 后面的句子有词语省略，补全应为：
… the voltage across the reactors is the full system voltage at 90 degrees firing angle and is zero at 180 degrees firing angle.

**3. The capacitor switching (connecting) takes place at that specific instant in a cycle at which the conditions for minimum transients are satisfied, that is, when the voltage across the thyristor valve is zero or minimum.**

译文：电容投切（连接）发生在一个周期中的特定时刻，此时的最小暂态现象令人满意，即，晶闸管阀上的压降为零或最小的时候。

说明：at which 引导的定语从句修饰 specific instant。at that instant at which 与 when 的含义接近，都表示发生的时刻，引导时间状语从句，这两个状语从句是并列关系。

**4. The response time of SVCs is in the range of 2-3 cycles which makes SVCs well suited for applications in which fast and repetitive control of reactive power is required.**

译文：SVC 的响应时间在 2-3 个周期的范围内，使其很适合用在需要无功功率的快速、重复控制的应用中。

# Chapter 21　Flexible AC Transmission System

说明：in the range of 表示"在…范围内"。which 和 in which 引导的定语从句，分别修饰 the range of 2-3 cycles 和 applications。

**5. The Static Synchronous Compensator (STATCOM) is a principal state-of-the-art FACTS equipment and is now a commercially available additional tool for use by system planners and designers for shunt reactive power compensation at either the transmission level for voltage regulation and system stability improvement or at the distribution level for voltage regulation, power factor correction and/or improvement of power quality.**

译文：静止同步补偿器（STATCOM）是重要的、达到最新发展水平的 FACTS 设备，现在作为可用的额外工具被系统规划和设计人员用于并联无功补偿，要么在输电级用于电压调整和系统稳定性改善，要么在配电级用于电压调整、功率因数校正和/或电能质量的改善。

说明：这是一个典型的英文长句。翻译是要注意整个句子的通顺流畅。state-of-the-art 意思是"达到最新技术发展水平的"。句式表示"要么…，要么…"，或"或者…或者…"。

**6. As with all static FACTS devices the STATCOM has the potential to be exceptionally reliable but with the added capability to ...**

译文：STATCOM 在和所有静态 FACTS 设备一样具有异常可靠的潜能的同时，此外还有……的能力。

说明：句首的 As with 表示"和……一样，与……一致"。but 表示"除……之外，此外"。注意：千万不能把 with all 看作一体，译为"尽管"，更不能把 but 译为"但是"。

## ABBREVIATIONS (ABBR.)

| | | | |
|---|---|---|---|
| 1. | FACTS | flexible aC transmission system | 柔性交流输电系统 |
| 2. | GTO | gate turn-off thyristor | 门极可关断晶闸管 |
| 3. | IGBT | insulated gate bipolar transistor | 绝缘栅型双极性晶体管 |
| 4. | IGCT | integrated gate commutation transistor | 集成门极可换向晶体管 |
| 5. | SVC | static var compensator | 静止无功补偿器 |
| 6. | SVG | static var generator | 静止无功发生器 |
| 7. | TCR | thyristor-controlled reactor | 晶闸管可控电抗器 |
| 8. | TSR | thyristor-switched reactor | 晶闸管投切电抗器 |
| 9. | TSC | thyristor-switched capacitor | 晶闸管投切电容器 |
| 10. | TCT | thyristor-controlled transformer | 晶闸管可控变压器 |
| 11. | SCT | self- commutated transformer | 自换向变压器 |
| 12. | LCT | line- commutated transformer | 线换向变压器 |
| 13. | SCC | self- commutated converter | 自换向换流器 |
| 14. | LCC | line-commutated converter | 线换向换流器 |
| 15. | TCSC | thyristor controlled series compensator | 晶闸管可控串联补偿 |
| 16. | VSC | voltage source converter | 电压源换流器 |
| 17. | VSI | voltage source inverter | 电压源逆变器 |
| 18. | STATCOM | static synchronous compensator | 静止同步补偿器 |

| 19. | PWM | pulse width modulation | 脉（冲）宽（度）调制 |
| 20. | DVR | dynamic voltage restorer | 动态电压恢复器 |
| 21. | APF | active power filters | 有源电力滤波器 |

# EXERCISES

**1. Translate the following words or expressions into Chinese.**

(1) FACTS  (2) transistor  (3) firing pulse
(4) square-wave  (5) energy storage  (6) state-of-the-art
(7) solid-state  (8) footprint  (9) Thevenin equivalent

**2. Translate the following words or expressions into English.**

(1) 电力电子  (2) 二极管  (3) 动态电压恢复器
(4) 分步的，有级的  (5) 换向的  (6) 调谐

**3. Fill in the blanks with proper words or expressions.**

(1) The compensator supplies reactive power to the system when it is operated in the _____ mode. In contrast, when it operates in the _____ mode, the STATCOM absorbs reactive power from the systems.

(2) A dynamic voltage restorer (DVR) is a device that generates an AC voltage and injects it in _____ with the supply voltage through an injection transformer to compensate for the voltage _____.

(3) The circuit configurations of APFs are similar to that of STATCOMs, which consist of an _____ based on power electronic switching elements and converts the _____ voltage on a capacitor to _____ voltage on the system side.

(4) A Static Var Compensator (SVC), according to the basic terminology established by CIGRE, is a _____-connected static var _____ (SVG) whose output is varied so as to maintain or control specific parameters of the electric power system.

# Word-Building (20)　bi-, di-; tri-　双，二；三

## 1. bi-[ 前缀 ]，表示：二，两，双

| bilateral | adj. | 双向的，双面的，双边的 |
| bidirectional | adj. | 双向的 |
| bimotored | adj. | 双发动机的 |
| bistable | adj. | 双稳（态）的 |
| biphase | adj.n. | 双相（的） |
| bipolar | adj. | 双极的，有两极的 |

# Chapter 21　Flexible AC Transmission System

| bicycle | n. | 两轮车，自行车 |

## 2. di-[前缀]，表示：双，偶，两

| diode | n. | 二极管 |
| dipole | n. | 双极子，偶极 |
| dioxide | n. | 二氧化物 |

## 3. tri-[前缀]，表示：三,三重,三次,三倍

| triode | n. | 三极真空管 |
| triangle | n. | 三角形 |
| triangular | adj. | 三角形的，三人间的 |
| triband | n. | 三重频带 |
| tricycle | n. | 三轮车，机器三轮车 |
| triaxial | adj. | 三轴的，三维的，空间的 |
| tridimensional | adj. | 三度空间的，立体的 |

# Chapter 22
# Power Quality and Its Improvement

## Part 1 Familiar Power Quality Problems

Power quality is the combination of voltage quality and current quality. Thus power quality is concerned with deviations of voltage and/or current from the ideal. Note that power quality has nothing to do with deviations of the product of voltage and current (the power) from any ideal shape.*

Power quality has during recent years achieved an increasing interest. The concept of power quality includes the quality of the supplying voltage with respect to, for instance, interruptions, voltage dips, voltage sag/swell, imbalance, harmonic distortion, voltage fluctuation and flicker, and some kinds of high-frequency transients, etc.*

Series and shunt active power conditioners have been proposed and used for compensating purposes.

Before further description of definite power quality problems, it is necessary to introduce some relative concepts.

A point of common coupling (PCC) is a very common item in power quality studies. By definition, a PCC is the point of the public supply network, electrically nearest to a particular consumer installations, and at which other consumer installations are, or may be, connected.* The feeders terminate at certain PCCs.

As for describing the degree of distortion of a current or voltage, there are several ways. Two that are frequently used are crest factor and total harmonic distortion (THD).

By definition, the crest factor of a voltage is equal to the peak value divided by the effective value also called the root mean square (RMS) value

$$\text{crest factor} = \frac{\text{peak voltage}}{\text{effective voltage}}$$

In the case of a sinusoidal voltage (which has no distortion), the crest factor is $\sqrt{2}$. A wave having a crest factor less than $\sqrt{2}$ tends to be flat-topped. On the other hand, a crest factor greater than $\sqrt{2}$ indicates a voltage that tends to be pointy.

# Chapter 22  Power Quality and Its Improvement

## Part 2  Harmonics and Nonlinear Loads

The voltages and currents in a power circuit are frequently not pure sine waves. The line voltages usually have a satisfactory wave shape but the currents are sometimes badly distorted, as shown in Figure 22.1. This distortion can be produced, for instance, by magnetic saturation in the cores of transformers or by the switching action of thyristors or IGBTs in electronic drives.

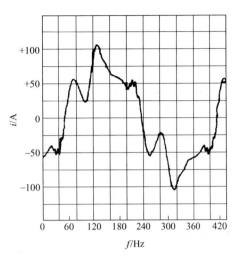

**Figure 22.1**  A distorted current

The distortion of a voltage or current can be traced to the harmonics it contains. A harmonic is any voltage or current whose frequency is an integral multiple of (2, 3, 4, etc, times) the line frequency.

Consider a set of sine waves in which the lowest frequency is $f$, and all other frequencies are integral multiples of $f$. By definition, the sine wave having the lowest frequency is called the *fundamental*, and the other waves are called *harmonics*. For example, a set of sine waves whose frequencies are 20, 40, 100, and 380Hz is said to possess the following components:

| | | |
|---|---|---|
| fundamental: | 20Hz | (the lowest frequency) |
| second harmonic: | 40Hz | (2 × 20Hz) |
| fifth harmonic: | 100Hz | (5 × 20Hz) |
| nineteenth harmonic: | 380Hz | (19 × 20Hz) |

Thus, the sum of a fundamental voltage and a harmonic voltage yields a non-sinusoidal waveform whose degree of distortion depends upon the magnitude of the harmonic (or harmonics) it contains.

We can produce a periodic voltage or current of any conceivable shape. All we have to do is to add together a fundamental component and an arbitrary set of harmonic components.

A square wave is thus composed of a fundamental wave and an infinite number of harmonics. The higher harmonics have smaller and smaller amplitudes, and they are consequently less important. However, these high-frequency harmonics produce the steep sides and pointy corners of the square wave. In practice, square waves are not produced by adding sine waves, but the example does show that any waveshape can be built up from a fundamental wave and an appropriate number of harmonics.

Conversely, we can decompose a distorted periodic wave into its fundamental and harmonic components.

Harmonic voltages and currents are usually undesirable, but in some AC circuits, they are also unavoidable.

The widespread use of nonlinear loads is leading to a variety of undesirable phenomena in the operation of power systems. The harmonic components in current and voltage waveforms are the most important among these. Many harmonics are created by nonlinear loads, such as electric arcs and saturated magnetic circuits. They are also produced whenever voltages and currents are periodically switched, such as in power electronic circuits. All these circuits produce distorted waveshapes that are rich in harmonics.

Consider a sinusoidal voltage connected to a nonlinear load. The load may be a saturable reactance, a rectifier, or a set of mechanical switches that open and close periodically. On account of the nonlinearity, the current will not be sinusoidal. It will contain a fundamental component and harmonic components. The fundamental component is produced by the sinusoidal voltage, but the harmonic components are generated by the load. The harmonic currents flow in the sinusoidal source as well as in the load.

As far as the fundamental component of current is concerned, it can lag, lead, or be in phase with source voltage. Thus, we can attribute traditional active and reactive powers to this non-linear load.

In AC circuits, the fundamental current and fundamental voltage together produce fundamental power. This fundamental power is the useful power that causes a motor to rotate and a furnace to heat up. The product of a harmonic voltage times the corresponding harmonic current also produces a harmonic power.* The latter is usually dissipated as heat in the AC circuit and, consequently, does no useful work. Harmonic currents and voltages should therefore be kept as small as possible. It should be noted that the product of a fundamental voltage and a harmonic current yields zero net power.

Passive filters have been used to eliminate line current harmonics. However, they introduce resonance in the power system and tend to be bulky. With the improved performance of power and control circuits, active power filters (APFs) have gradually been recognized as a viable alternative to passive filters.

## Part 3  Voltage Fluctuation and Flicker

There are many devices in the electric power system that impose rapid and frequent changes of load, with correspondingly rapid voltage changes such as voltage fluctuations and flickers.* Some of the most common devices are induction and synchronous motors, motor-driven reciprocating loads, motor-driven intermittent loads, electric arc furnaces (EAF), electric welders and miscellaneous installations such as electric shovels and rolling mills. The rapid and frequent voltage changes, especially flickers, caused by these devices can be perceived by and be objectionable to nearby customers and often result in complaints to the power companies.*

Flicker is understood to be the sensation that is experienced by humans when subjected to changes in the illumination intensity. The human maximum sensitivity to illumination changes is in the frequency range from about 5 to 15Hz. The fluctuating illumination is caused by amplitude modulation of the feeding alternating voltage.

# Chapter 22　Power Quality and Its Improvement

　　To be able to measure the level of flicker, a meter is needed. There are different standards existing describing flicker meters. The flicker meter used the most often is the one that is described by the International Electrotechnical Commission (IEC), sometimes referred to as the PST meter. "ST" stands for a short time, which in this case refers to measurements during 10 minutes. Unity output from the PST meter corresponds to a flicker level causing 50% of the persons in a reference group to be disturbed by the flicker.

　　Large industrial loads, such as EAFs, used for melting, for example, scrap with electric energy, cause voltage distortion like harmonics and voltage fluctuation in the feeding AC system. An arc furnace is probably one of the largest existing end-users of electric energy with a rated power up to the order of 100MVA. To limit the effects of these disturbing loads, compensation devices have usually to be connected.

　　Many corrective pieces of equipment and procedures have been used to minimize voltage flicker. The following conventional remedial measures are commonly considered for applications in distribution systems:
- Motor starters/ Adjustable speed drives (ASD);
- Shunt capacitors;
- Voltage regulators;
- Synchronous condensers;
- System changes.

　　Only some of the conventional corrective measures discussed above are effective in controlling voltage flickers. The ASD is not a generalized solution. The alternatives of synchronous condensers and system changes can be effective but generally are expensive and/or impractical. Utilities sometimes refuse to add new customers with equipment which will cause severe voltage flickers on the electric system because no effective and/or economical solution can be found.

　　With the advent of compact FACTS devices, distribution engineers have another viable option available to solve many of the voltage fluctuation problems in their systems.

　　The advantages of compact FACTS devices are summarized:
- Fast voltage control which can be repeated frequently;
- Small physical size;
- Flexibility of location;
- Comparable economics.

## Part 4　Unbalanced Loads

　　Electric tractions are typical unbalanced loads. Voltage and current unbalance between phases of AC supply systems must likewise be confined in magnitude and prevented from spreading through the grid into other parts of the system, lest they become a nuisance to others.*

There are a number of different ways to feed traction systems with electric power. One modern system used in many recent electrification projects is to directly supply it with 50Hz mains power. The transmission/sub-transmission voltages are then directly transformed by a power transformer to the traction voltage. There are two competing systems on the traction side, the booster transformer scheme and the autotransformer scheme. In the booster transformer scheme (Figure 22.2), the mains voltage is transformed into one single-phase voltage of 25kV. One of the power transformer traction winding ends is earthed and the other end is connected to the catenary wire. In the autotransformer scheme (Figure 22.3), the traction winding is connected to the earth at its midpoint. The other two ends of the winding are connected to the catenary wire and the feeder wire respectively. The earthed points are connected to the rail in both schemes.

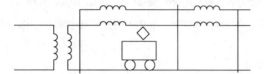 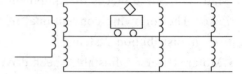

**Figure 22.2**  the booster transformer scheme       **Figure 22.3**  the auto transformer scheme

On the transmission network side, the power transformer is connected between two phases. Frequently, two isolated rail sections are fed from the same feeder station. In this case, the power transformers are connected between different phases. The traction load is relatively large; today it is common with power ratings in the range of 50-100MW per feeding transformer. These loads connected between two phases on the mains will create unbalances in the supply system voltage.

A common requirement is that the negative phase sequence voltage resulting from an unbalanced load should not exceed 1%. In many cases, the traction system is relatively far apart from strong high-voltage transmission lines, while weaker sub-transmission lines normally run somewhere in the vicinity of the rail. These lines can be utilized for the rail supply in case the unbalance caused by the traction load can be eliminated/mitigated. There are means available today by controllable high voltage power electronic equipment for unbalanced compensation/suppression. Conventional SVCs or the most recently developed D-STATCOM (called SVC-Light by some authors) based on voltage source converters can serve as "load balancers" by use of special control algorithms.

The term "load compensation" means to balance the unbalanced load and correct the load power factor to unity at the same time. This is very important for some applications such as the compensations of single-phase railway systems and EAF systems.

The method of symmetrical components is used for deriving the compensation scheme.

For the fast load compensation, the compensator should compensate the imaginary part of the positive-sequence load current and the entire negative-sequence load current as soon as possible. In this way, the power source supplies only a real part of the positive-sequence load current. Since no zero-sequence component appears in a three-phase three-wire system, the compensation current can be derived. The D-STATCOM is here treated as a current-controlled source to supply the needed compensation current.

# Part 5　Voltage Sag

Voltage magnitude is one of the major factors that determine the quality of power supply. Loads at the distribution level are usually subject to frequent voltage sags due to various reasons. Voltage sags are highly undesirable for some sensitive loads, especially in high-tech industries. It is a challenging task to correct the voltage sag so that the desired load voltage magnitude can be maintained during the voltage disturbances.

DVR or D-STATCOM can be used to correct the voltage sag at the distribution level. A DVR is a series device that generates an AC voltage and injects it in series with the supply voltage through an injection transformer to compensate for the voltage sag. The injected voltage and load current determine the power injection of the DVR. On the other hand, a D-STATCOM is a shunt device that generates an AC voltage, which in turn causes a current injection into the system trough a shunt transformer. The load voltage and injected current determine the power injection of the D-STATCOM. For lower voltage sags, the load voltage magnitude can be corrected by injecting only reactive power into the system. However, for higher voltage sags, injection of active power, in addition to reactive power, is essential to correct the voltage magnitude. Note that both DVR and D-STATCOM are capable of generating or absorbing reactive power but the active power injection of the device must be provided by an external energy source or energy storage system.

The response time of both DVR and D-STATCOM is very short and is limited by the power electronics devices and the voltage sag detection time. The expected response time is about 25ms, is much less than some traditional methods of voltage correction such as tap-changing transformers.

## NEW WORDS AND EXPRESSIONS

| | | | |
|---|---|---|---|
| 1. | power quality | | 电能质量 |
| 2. | dip | n. | 跌落，下垂 |
| 3. | swell | n. | 凸起，暂升 |
| 4. | distortion | n. | 畸变 |
| 5. | fluctuation | n. | 波动，起伏 |
| 6. | flicker | n. | 闪变 |
| 7. | conditioner | n. | 调节装置 |
| 8. | crest factor | n. | 波峰因数（振幅与有效值之比） |
| 9. | evidently | adv | 明显地，显然 |
| 10. | flat-topped | | 平顶的 |
| 11. | pointy | adj. | 非常尖的 |
| 12. | integral multiple | | 整倍数 |

| | | | |
|---|---|---|---|
| 13. | fundamental | n.adj. | 基频（的），基波（的） |
| 14. | conceivable | adj. | 可能的，想得到的 |
| 15. | arbitrary | adj. | 任意的，专断的 |
| 16. | square wave | | 方波 |
| 17. | steep side | | 陡坡，陡沿 |
| 18. | pointy corner | | 尖角 |
| 19. | decompose | vt. | 分解 |
| 20. | magnetic circuit | | 磁路 |
| 21. | saturable | adj. | 可饱和的 |
| 22. | furnace | n. | 炉子，熔炉 |
| 23. | dissipate | v. | 散失，消耗 |
| 24. | useful work | | 有用功 |
| 25. | net power | | 净[有效]功率 |
| 26. | viable alternative | | 可行的替换物 |
| 27. | reciprocating | adj. | 往复的，来回的 |
| 28. | intermittent load | | 间歇[断续]荷载 |
| 29. | welder | n. | 电焊，焊接装置 |
| 30. | miscellaneous | adj. | 各种各样的，混杂的 |
| 31. | shovel | n. | 铲，铁铲 |
| 32. | rolling mill | | 轧钢厂 |
| 33. | sensation | n. | 感觉 |
| 34. | illumination | n. | 照度，发光强度 |
| 35. | unity | n. | 单位1 |
| 36. | scrap | n. | 废料 |
| 37. | remedial measure | | 补救措施 |
| 38. | advent | n. | 出现，到来 |
| 39. | compact | adj. | 紧凑的 |
| 40. | electric traction | | 电力牵引 |
| 41. | lest | con | 唯恐，以免，免得 |
| 42. | electrification | n. | 电气化 |
| 43. | booster | n. | 升降压器 |
| 44. | catenary wire | | 悬链线 |
| 45. | mitigate | v. | 减轻 |
| 46. | imaginary par | | （复数的）虚部 |

## PHRASES

**1. be concerned with:** 常表示"参与，干预"，在此表示"关注，对…感兴趣"

Example: Thus power quality <u>is concerned with</u> deviations of voltage and/or current from the ideal.

因而电能质量关注电压和/或电流与理想情况的偏离。

# Chapter 22　Power Quality and Its Improvement

**2. trace to: 上溯**

Example: The distortion of a voltage or current can be traced to the harmonics it contains.

电压或电流的畸变可以追溯到其所包含的谐波。

**3. a variety of: 多种的**

Example: The widespread use of nonlinear loads is leading to a variety of undesirable phenomena in the operation of power systems.

非线性负荷的广泛使用导致电力系统运行中的多种不理想的现象。

**4. on account of: 由于**

Example: On account of the nonlinearity, the current will not be sinusoidal.

由于非线性，电流不会是正弦的。

**5. attribute to: 归因于（在本例中可译为"应用于"）**

Example: Thus, we can attribute traditional active and reactive powers to this non-linear load.

于是，我们可以把传统的有功、无功功率应用于该非线性负荷。

## COMPLICATED SENTENCES

**1. Note that power quality has nothing to do with deviations of the product of voltage and current (the power) from any ideal shape.**

译文：注意，电能质量和电压、电流乘积（功率）与任何理想形态的偏差无关。

说明：has nothing to do with 表示"与……无关"。product 意思是"乘积"，在此不能译为"产生"。deviation from ideal shape 表示"与理想形态的偏差"。

**2. The concept of power quality includes the quality of the supplying voltage with respect to, for instance, interruptions, voltage dips, voltage sag/swell, imbalance, harmonic distortion, voltage fluctuation and flicker, and some kinds of high frequency transients, etc.**

译文：电能质量的概念包括供电电压质量，例如，关于断电、电压跌落、电压暂降与暂升、不平衡、谐波畸变、电压波动与闪变以及某些种类的高频暂态现象，等等。

说明：句子的主体结构是 The concept of power quality includes the quality of the supplying voltage with respect to，意思是"电能质量的概念包括关于……的供电电压质量"。with respect to 表示"关于"，在翻译时为了句子通顺，可以隐含于句中，不必明译。for instance 意思是"例如"。

**3. By definition, a PCC is the point of the public supply network, electrically nearest to a particular consumer installations, and at which other consumer installations are, or may be, connected.**

译文：按照定义，PCC 是公共供电网络在电气上距某一特定用户设备最近的点，其他用户的设备也连接或者可以连接到那里。

说明：at which 引导的定语从句修饰 the point。在该定语从句中，谓语部分有省略，补全应为：other consumer's installation are connected or may be connected。

**4. The product of a harmonic voltage times the corresponding harmonic current also produces a harmonic power.**

译文：谐波电压乘以对应谐波电流的积也产生一个谐波功率。

说明：product 意思是"乘积"，times 表示"乘以，与……相乘"。这种用法在英文科技文献中比较常见，应重点掌握。

**5. There are many devices in the electric power system that impose rapid and frequent changes of load, with correspondingly rapid voltage changes such as voltage fluctuations and flickers.**

译文：电力系统中有很多造成负荷快速而频繁变化的设备，这些负荷变化对应着电压波动和闪变等快速的电压变化。

说明：which 引导的定语从句修饰的主体是 devices，而从 with 开始的介宾短语修饰 impose rapid and frequent changes of load。在翻译时，为了句子通顺，可以做适当的语序调整。

**6. The rapid and frequent voltage changes, especially flickers, caused by these devices can be perceived by and be objectionable to nearby customers and often result in complaints to the power companies.**

译文：这些设备引起的快速而频繁的电压变化，尤其是闪变，能被附近用户感知而且比较令人讨厌，常常导致对电力公司的抱怨。

说明：be perceived by and be objectionable to nearby customers 中，由于介词 by 和 to 具有相同的宾语，在 by 的后面省略了 nearby customers。补全应为：be perceived by nearby customers and be objectionable to nearby customers。当然，在翻译时也可以做适当的省略。

**7. Voltage and current unbalances between phases of AC supply systems must likewise be confined in magnitude and prevented from spreading through the grid into other parts of the system, lest they become a nuisance to others.**

译文：交流供电系统相间的电压、电流不平衡，同样必须限制其大小并防止通过电网扩散到系统的其他部分，以免它们成为别人的麻烦。

说明：likewise 表示"同样地，也"，连词 lest 表示"唯恐，以免，免得"。prevent from spreading 意思是"唯恐，以免，免得"。在这里，the grid 指的是"电网"。

## ABBREVIATIONS (ABBR.)

| | | | |
|---|---|---|---|
| 1. | PCC | point of common coupling | 公共联结点 |
| 2. | THD | total harmonic distortion | 总谐波畸变（失真） |
| 3. | rms | root mean square | 均方根 |
| 4. | EAF | electric arc furnaces | 电弧炉 |
| 5. | IEC | international electrotechnical commission | 国际电工委员会 |

# Chapter 22  Power Quality and Its Improvement

    6.    ASD    adjustable speed drive    调速驱动器

## SUMMARY OF GLOSSARY

1. 电能质量问题　power quality problems
   - imbalance　　　　　　　　　　　　　不平衡
   - interruption　　　　　　　　　　　　　供电中断
   - voltage dip　　　　　　　　　　　　　电压跌落
   - voltage sag/swell　　　　　　　　　　电压暂降/暂升
   - voltage fluctuation/flicker　　　　　　电压波动与闪变
   - harmonic distortion　　　　　　　　　谐波畸变
   - high frequency transient　　　　　　　高频暂态现象
2. 波形关键数值　key values of a waveform
   - peak value　　　　　　　　　　　　　峰值
   - effective value　　　　　　　　　　　有效值
   - rms value　　　　　　　　　　　　　均方根值
3. 畸变波形成分　components of distorted waveform
   - fundamental　　　　　　　　　　　　基波
   - harmonics　　　　　　　　　　　　　谐波
4. 相位关系　phase relationship
   - lag　　　　　　　　　　　　　　　　领先，超前
   - lead　　　　　　　　　　　　　　　　滞后，落后
   - be in phase with　　　　　　　　　　与……同相
5. 滤波器　filter
   - passive filter　　　　　　　　　　　　无源滤波器
   - active power filter　　　　　　　　　　有源电力滤波器
6. 对称分量法　method of symmetrical components
   - zero-sequence component　　　　　　零序分量
   - positive-sequence component　　　　　正序分量
   - negative-sequence component　　　　　负序分量
7. 复数　complex number
   - real part　　　　　　　　　　　　　　实部
   - imaginary part　　　　　　　　　　　虚部

# EXERCISES

**1. Translate the following words or expressions into Chinese.**

(1) harmonic distortion　　(2) fluctuation　　(3) voltage sag
(4) crest factor　　(5) square wave　　(6) net power
(7) resonance　　(8) unity　　(9) booster

**2. Translate the following words or expressions into English.**

(1) 电能质量　　(2) 闪变　　(3) 基频

(4) 有用功　　　　　　(5) 对称分量法　　　　　　(6) 电力牵引

**3. Fill in the blanks with proper words or expressions.**

(1) Power quality is concerned with deviations of _____ and/or _____ from the ideal, while the power quality has nothing to do with deviations of the product of them (the _____) from any ideal shape.

(2) Consider a sinusoidal voltage connected to a nonlinear load. On account of the non-linearity, the current will not be sinusoidal. It will contain a _____ component produced by the sinusoidal voltage and _____ components generated by the load.

(3) The term "load compensation" means to balance the _____ load and correct the load _____ to unity at the same time.

(4) Note that both DVR and D-STATCOM are capable of generating or absorbing _____ power but the _____ power injection of the device must be provided by an external energy source or _____ system.

# Word-Building (21)　de-, anti-; counter-　反，抗，逆

### 1. de-[前缀]，表示相反动作，剥夺，分离：分，解

| | | | |
|---|---|---|---|
| code | decode | vt. | 解码，译解 |
| compose | decompose | v. | （使）分解 |
| form | deform | v. | （使）变形 |
| modulate | demodulate | vt | （使）解调 |
| struction | destruction | n. | 破坏，毁灭 |

### 2. anti-[前缀]，表示"反对，抵抗"之义：反、抗、防、逆、耐

| | | | |
|---|---|---|---|
| aging | antiaging | adj. | 防衰老的 |
| magnetic | antimagnetic | adj. | 防磁的 |
| interference | anti-interference | a.n. | 抗干扰（的） |
| jam | antijam | v.n. | 抗干扰 |
| virus | antivirus | n. | 抗病毒，防病毒 |
| phase | antiphase | n. | 反相 |
| bouncer | antibouncer | n. | 减振器，防回跳装置 |
| coincidence | anticoincidence | n. | 反重合，非一致，异 |
| parallel | antiparallel | adj. | 反平行的，反并联的 |

### 3. counter-[前缀]，表示"相反，相对"之义：反，逆

| | | | |
|---|---|---|---|
| clockwise | counterclockwise | ad. | 逆时针，反时针方向（的） |
| current | countercurrent | n. | 逆流，[电]反向电流 |
| device | counterdevice | n. | 对抗装置 |
| flow | counterflow | n. | 逆流 |
| force | counterforce | n. | 反作用力 |

# Chapter 23
# Electromagnetic Field and Electromagnetic Compatibility

## Part 1  Overview of Electromagnetic Field

People use magnets to make generators but also use electricity to make electromagnets. The discovery of electricity and magnetism is a gradual process, which is the crystallization of the sweat of many scientists. Coulomb, a French physicist, first studied the force between charges and proposed Coulomb's law. Since then, scientists have been debating the way of the force between charges: some people believe that the force between charges does not require time and space, and one charge will exert a force on another in a moment, which is called "action at a distance". With the development of science, British scientist Faraday put forward the concept of the "electric field".

Austrian, a Danish physicist, accidentally discovered the phenomenon of electricity-generating magnetism in a classroom experiment. Ampere pointed out the judgment method of the direction of the magnetic field generated by the current: the right-hand spiral rule. Under the inspiration of electricity-generating magnetism, Faraday searched for the possibility of magnetism generating electricity and found the phenomenon of electromagnetic induction. To understand electromagnetic fields, there are several related concepts:

1) Gradient: In vector calculus, the gradient of a scalar field is a vector field. The gradient at a certain point in the scalar field points to the direction where the scalar field grows fastest, and the length of the gradient is the maximum rate of change.

2) Divergence: Divergence refers to the rate of change per unit volume when a fluid moves.

3) Curl: The amount by which a curve, fluid, etc. is rotated.

## Part 2  Maxwell's Equations

Maxwell's equations consist of four equations, all of which have differential and integral expressions. The four equations are Gaussian law, Gaussian magnetic law, full current law, and electromagnetic induction law.

Gauss's law reveals how the electric charge generates an electric field. The electric field is active, and its source is the charge in space. The first viewpoint of Faraday is expressed in mathematical form: charge will generate an electric field in the surrounding space. Positive charges will emit electric field lines, and negative charges will absorb electric field lines from the surroundings. The greater the quantity of electric charge, the more electric field lines are emitted or absorbed. By calculating the number of electric field lines passing through a given closed surface, namely its electric flux, we can know the total charge contained in the closed surface. More specifically, this law describes the relationship between the electric flux passing through an arbitrary closed surface and the electric charge in the closed surface. ( $\nabla \cdot E = \dfrac{\rho}{\epsilon_0}$ )

Gauss's magnetic law reveals the continuity of magnetic flux. Different from an electric field, the magnetic induction line is always closed whether it is the magnetic field generated by magnets or the magnetic field generated by current. The magnetic induction line has neither a starting point nor an ending point. The magnetic flux passing through any closed surface is equal to zero, or the magnetic field is a passive field. ( $\nabla \cdot B = 0$ )

The law of electromagnetic induction mathematically explains the origin of Faraday's law of electromagnetic induction, which can also be described as a swirling field. When a magnet is close to a coil, an induced current is generated in the coil. The curl of the electric field strength E is equal to the negative value of the time change rate of the magnetic flux density B at this point, that is, the vortex source of the electric field is the time change rate of the magnetic flux density. ( $\nabla \times E = -\dfrac{\partial B}{\partial t}$ )

The law of full current reveals the cause of the magnetic field, which shows that the curl of magnetic field strength H is equal to the full current density of the point. ( $\nabla \times H = J + \dfrac{\partial D}{\partial t}$ )

Maxwell's equations tell us that electric field and magnetic field do not exist alone, but are unified in an electromagnetic field. Not only that, Maxwell also proved through calculation that if there is an oscillating electric field in the vacuum, then a magnetic field will be generated around the oscillating electric field, and this magnetic field will further generate an electric field... So to and fro, the electromagnetic field can spread far away, forming electromagnetic waves.

## Part 3  Electromagnetic Compatibility

Electromagnetic compatibility (EMC) is defined as the ability of a device, equipment, or system to function satisfactorily in its electromagnetic environment without introducing intolerable electromagnetic disturbances to anything in that environment.

# Chapter 23  Electromagnetic Field and Electromagnetic Compatibility

There are two aspects to EMC: ① a piece of equipment should be able to operate normally in its environment, and ② it should not pollute the environment too much. In EMC terms: degradation and emission. The third term of importance is the "electromagnetic environment", which gives the level of disturbance against which the equipment should be immune. Within the EMC standards, a distinction is made between radiated disturbances and conducted disturbances. Radiated disturbances are emitted (transmitted) by one device and received by another without the need for any conduction. Conducted disturbances need a conductor to transfer from one device to another. These conducted disturbances are within the scope of power quality; radiated disturbances (although very important) are outside of the normal realm of power system engineering or power quality.

A schematic overview of the EMC terminology is given in Figure 23.1. A special type of disturbance, not shown in the figure, is radiated disturbances which induce conducted disturbances in the power system.

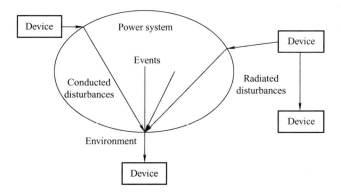

**Figure 23.1**  Overview of EMC Terminology

### 1. Immunity requirements

Immunity standards define the minimum level of electromagnetic disturbance that a piece of equipment shall be able to withstand. In practice it will often be clear when a device performs satisfactorily and when not, but when testing equipment, the distinction may become blurred. It will all depend on the application whether or not a certain equipment behavior is acceptable.

The basic immunity standard gives four classes of equipment performance:
- Normal performance within the specification limits.
- Temporary degradation or loss of function which is self-recoverable.
- Temporary degradation or loss of function which requires operator intervention or system reset.
- Degradation or loss of function that is not recoverable due to damage of equipment, components, or software, or loss of data.

These classes are general as the description should apply to all kinds of equipment. This classification is further defined in the various equipment standards.

## 2. Emission Standards

Emission standards define the maximum amount of electromagnetic disturbance that a piece of equipment is allowed to produce. Within the existing IEC (International Electrotechnical Commission) standards, emission limits exist for harmonic currents and for voltage fluctuations. Most power quality phenomena are not due to equipment emission but due to operational actions or faults in the power system. As the EMC standards only apply to equipment, there are no "emission limits" for the power system. Events like voltage sags and interruptions are considered a "fact of life". These events do, however, contribute to the electromagnetic environment.

## 3. The electromagnetic environment

To give quantitative levels for the immunity of equipment, the electromagnetic environment should be known. The electromagnetic environment for disturbances originating in or conducted through the power system is equivalent to the voltage quality.* The IEC electromagnetic compatibility standards define the voltage quality in three ways:

1) Capability levels are reference values for coordinating emission and immunity requirements of equipment. For a given disturbance, the compatibility level is in between the emission level (or the environment) and the immunity level. As both emission and immunity are stochastic quantities, EMC can never be completely guaranteed. The compatibility level is chosen such that compatibility is achieved for most equipment most of the time: typically, 95% of equipment, 95% of the time.*

2) Voltage characteristics are quasi-guaranteed limits for some parameters, covering any location. Again, the voltage characteristics are based on a 95% value, but now only in time. They hold at any location and are thus an important parameter for the customer. Voltage characteristics are a way of describing electricity as a product.

3) Planning levels are specified by the supply utility and can be considered as internal quality objectives of the utility.

These ideas were originally developed for disturbances generated by equipment, for which other equipment could be sensitive: mainly radio frequency interference. These ideas have been extended towards variations like harmonic distortion or voltage fluctuations. The concept has not yet been applied successfully to events like voltage sags or interruptions.

# Part 4  Design on EMC in Devices

There is Electromagnetic interference (EMI) exists between electronic or electrical devices and communication devices. Since the power network is exposed to the natural environment, the EMI may result in some undesirable problems as follows:

1) maloperations of the control circuit in a device caused by external noise;

# Chapter 23　Electromagnetic Field and Electromagnetic Compatibility

2) maloperations of communication devices due to the noise caused by electrical devices;

3) noise on power network due to those in electrical devices;

4) noise emission into the environment from devices.

As mentioned above, on the one hand, there are often various circuits for drive, protection, program, and signal supervision in electrical devices, which consist of various integrated circuits and must be protected against interference by external noise. On the other hand, noise ingoing from the input terminal may appear in the output terminal, which as a transfer of noise must also be avoided. Therefore, EMC considerations or countermeasures of noise suppressing must be included in the design of electronic or electrical devices.

First, countermeasures of noise suppressing rest with the design of a noise filter. There are two kinds of noise transmitting, namely *common-mode noise* which flows into the earth, and *differential-mode noise* which flows in the circuit inside. As far as the countermeasures of noise suppressing are concerned, common-mode noise is of major interest. In the range of lower frequencies, however, a large proportion of noise problems are about differential-mode transmitting. Proper filters must be chosen according to the noise components. Among the noise filters, presently, most are dispersed noise filters in which the major elements are common-mode windings and line capacitors. The common-mode windings are used for damping common-mode noise. Normally, the windings are copper coils winded in the same direction wrapped around the ferrite core. At this rate, large inductance related to the magnetic saturation caused by alternating power of a 50Hz frequency in common-mode conditions may be prevented. Figure 23.2 is a sketch map of common-mode and differential-mode noise. Figure 23.3 demonstrates the principle of common-mode noise.

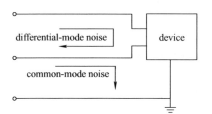

**Figure 23.2**　sketch map of common-mode and differential-mode noise

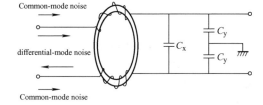

**Figure 23.3**　principle of common-mode noise

Surge protection is another important measure. Here surge protection is referred mainly to as lightning shielding, namely to the discharge of the pulse energy induced by lighting shocks in a very short time in order to protect the entire device. *At present, three means are used for surge protection, which are voltage-sensitive resistor, voltage-regulator diode and gas discharge tube. Fast response is the basic requirement for a suppressor; otherwise, harm will already come forth before the action of the suppressor if a fast-increasing instantaneous peak voltage appears across the protected circuit.*

Concerns with regard to EMC in the design of devices include:

(1) Electromagnetic Interference (EMI). EMI signals produced by the device under test

when in operation must be examined. EMI signals, also called noises, include conducted noise and radiated noise.

(2) Electromagnetic Sensitivity (EMS). This is aiming at the performance of anti-interference, which requires the device under test to have the ability to withstand the effect of some certain conducted or radiated disturbances without malfunctions or degradation of their capability.

(3) Electromagnetic Pulse (EMP). It's valuable to simulate the spikes coming forth on the power system caused by lightning and nuclear explosion and to impose them on devices under test to ensure no malfunction or degradation of capability will occur.* It is especially important to take measures to provide lightning shielding in the country.

(4) Electric Static Discharge (ESD). It's also important to simulate the static discharge characteristic and to impose the related voltage on devices under test to ensure no malfunction or degradation of capability will occur. This is based on the considerations of choice of elements, design of structure and circuit, and environment in which the devices are used.

## NEW WORDS AND EXPRESSIONS

| | | | |
|---|---|---|---|
| 1. | electromagnetic field | | 电磁场 |
| 2. | electromagnetic compatibility | | 电磁兼容 |
| 3. | charge | n. | 充电量，电荷 |
| 4. | Coulomb's law | | 库伦定律 |
| 5. | action at a distance | | 超距作用 |
| 6. | electric field | | 电场 |
| 7. | the right-hand spiral rule | | 右手螺旋定则 |
| 8. | electromagnetic induction | | 电磁感应 |
| 9. | divergence | n. | 散度 |
| 10. | curl | n. | 旋度 |
| 11. | Maxwell's Equations | | 麦克斯韦方程组 |
| 12. | Gaussian law | | 高斯定律 |
| 13. | Gaussian magnetic law | | 高斯磁定律 |
| 14. | full current law | | 全电流定律 |
| 15. | active field | | 有源场 |
| 16. | electromagnetic induction law | | 电磁感应定律 |
| 17. | magnetic flux | | 磁通量 |
| 18. | passive field | | 无源场 |
| 19. | swirling field | | 涡流场 |
| 20. | vortex | n. | 涡流 |
| 21. | oscillating | adj. | 振荡的 |
| 22. | vacuum | n. | 真空，真空状态 |
| 23. | immunity | n. | 抗扰度，免疫性 |
| 24. | emission | n. | 散发，发射 |
| 25. | immune | adj. | 免疫的 |
| 26. | emit | vt. | 发出，放射 |
| 27. | realm | n. | 领域 |

# Chapter 23  Electromagnetic Field and Electromagnetic Compatibility

| 28. | blurred | adj. | 模糊的 |
| 29. | degradation | n. | 降级，降格，退化 |
| 30. | intervention | n. | 干涉 |
| 31. | fact-of-life | | 无法更改的事实 |
| 32. | quantitative | adj. | 数量的，定量的 |
| 33. | stochastic | adj. | 随机的 |
| 34. | maloperation | n. | 误动作，不正确运行（维护） |
| 35. | integrated circuit | | 集成电路 |
| 36. | ingoing | n. | 进入 |
| 37. | countermeasure | n. | 对策，反措施 |
| 38. | common-mode | | 共模 |
| 39. | differential-mode | | 差模 |
| 40. | dispersed | adj. | 分散的 |
| 41. | ferrite core | | 铁氧体磁芯 |
| 42. | sketch map | | 草图，示意图 |
| 43. | surge protection | | 浪涌保护，过（高峰）电压保护 |
| 44. | suppressor | n. | 抑制器 |
| 45. | come forth | | 出现，涌现 |
| 46. | malfunction | n. | 故障 |
| 47. | spike | n. | 尖峰信号 |

## PHRASES

**1. So to and fro: 如此往复**

Example: So to and fro, the electromagnetic field can spread far away, forming electromagnetic waves.

如此往复，电磁场就可以向远处传播，形成电磁波。

**2. with reference to: 参照，也多用于表示"关于"**

Example: The immunity of the device should be assessed with reference to this electromagnetic environment.

设备的抗扰度应参照电磁环境进行评估。

**3. on the one hand...on the other hand...: 一方面…，另一方面…**

Example: As mentioned above, on the one hand, there are often various circuits for drive, protection, program, and signal supervision in electrical devices, which consist of various integrated circuits and must be protected against interference by external noise. On the other hand, noise ingoing from the input terminal may appear in the output terminal, which as a transfer of noise must also be avoided.

如上所述，一方面，电气设备中常常有用于驱动、保护、运行程序和信号监测的各种电路，它们由各种集成电路构成，必须保护其免受外来噪声干扰。另一方面，从输入端进入的噪声可能出现在输出端，必须避免该噪声传递。

**4. rest with: 取决于，在于，由…负责**

Example: Countermeasures of noise suppressing rest with the design of a noise filter.
噪声抑制对策在于噪声滤波器的设计。

**5. at this rate: 这样地话，这样地**

Example: At this rate, large inductance related to the magnetic saturation caused by alternating power of a 50Hz frequency in common-mode conditions may be prevented.
这样的话，可以防止 50Hz 交流电引起的磁通饱和在共模情况下获得大的电感

**6. with regard to: 关于**

Example: Concerns with regard to EMC in the design of devices include...
设备设计中有关 EMC 的事项包括……

## COMPLICATED SENTENCES

**1. The electromagnetic environment for disturbances originating in or conducted through the power system is equivalent to the voltage quality.**

译文：从电力系统发出的或通过电力系统传导的扰动，其电磁环境等同于电压质量。

说明：词组 is equivalent to 表示"相等[当]于…，等（同）于，与…等效"。originating in or conducted through the power system 是省略写法，补全应为：originating in the power system or conducted through the power system。

**2. The compatibility level is chosen such that compatibility is achieved for most equipment most of the time: typically 95% of equipment, 95% of the time.**

译文：兼容性水平选的是大多数设备在大部分时间里达到的兼容性：典型地取 95% 时间的 95% 的设备。

说明：词组 such that 表示"如此，这样"。在这里 equipment 是不可数名词。

**3. Here surge protection is referred mainly to as lightning shielding, namely to the discharge of the pulse energy induced by lighting shocks in a very short time in order to protect the entire device.**

译文：浪涌保护主要指防雷保护，就是在极短的时间内释放掉设备电流上因感应雷击而产生的大量脉冲能量到安全地线上，从而保护整个设备。

说明：be referred to 表示"是指，指的是"。namely 表示"即，也就是"

**4. Fast response is the basic requirement for a suppressor; otherwise, harm will already come forth before the action of the suppressor if a fast increasing instantaneous peak voltage appears across the protected circuit.**

译文：对抑制器的基本要求是反应速度要快，否则，如果在受保护电路的两端出现上升速度极快的瞬时峰值电压，那么在抑制器作用之前就会出现危害。

说明：come forth 意思是"出现，涌现"。

# Chapter 23　Electromagnetic Field and Electromagnetic Compatibility

**5. It's valuable to simulate the spikes coming forth on the power system caused by lightning and nuclear explosion and to impose them on devices under test to ensure no malfunction or degradation of capability will occur.**

译文：模拟因雷电和核爆炸在电网上生成的尖峰信号并施加在实验设备上，以保证不会导致设备的性能下降或故障，是有价值的。

说明：词组 impose on 表示"施加于"。under test 表示"测试中的，被测的"。

## ABBREVIATIONS (ABBR.)

1. EMC　　electro magnetic compatibility　　电磁兼容（性）
2. EMI　　electro magnetic interference　　电磁干扰
3. EMS　　electro magnetic sensitivity　　电磁敏感性
4. EMP　　electro magnetic pulse　　电磁脉冲
5. ESD　　electric static discharge　　静电放电
6. IEEE　　the Institute of Electrical and Electronics Engineers　　国际电气与电子工程师协会
7. Std　　standard　　标准
8. IEC　　International Electrotechnical Committee　　国际电工委员会
9. THD　　total harmonic distortion　　总谐波畸变

## SUMMARY OF GLOSSARY

1. 电磁兼容方面　aspects to EMC
   - immunity　　抗扰度，免疫性
   - emission　　散发，发射
   - electromagnetic environment　　电磁环境
2. EMC 标准中的扰动　disturbances in EMC standards
   - radiated disturbance　　辐射的扰动
   - conducted disturbance　　传导的扰动
3. 两种噪声　two kinds of noise
   - common-mode nois　　共模噪声
   - differential-mode noise　　差模噪声
4. 浪涌保护　surge protection
   - voltage-sensitive resistor　　压敏电阻
   - voltage-regulator diode　　稳压二极管
   - gas discharge tub　　气体放电管
5. 电磁场相关　electromagnetic field
   - electric field　　电场
   - magnetic field　　磁场
   - electromagnetic field　　电磁场
6. 向量计算　vector computation
   - gradient　　梯度
   - divergence　　旋度
   - curl　　散度

# EXERCISES

**1. Translate the following words or expressions into Chinese.**

(1) immunity  (2) degradation  (3) fact-of-life
(4) malfunction  (5) countermeasure  (6) common-mode
(7) stochastic

**2. Translate the following words or expressions into English.**

(1) 集成电路  (2) 差模噪声  (3) 铁氧体磁芯
(4) 浪涌保护  (5) 尖峰信号  (6) 误动作

**3. Fill in the blanks with proper words or expressions.**

(1) Electromagnetic _____ (EMC) is defined as the ability of a device, equipment or system to function satisfactorily in its _____ environment without introducing intolerable electromagnetic _____ to anything in that environment.

(2) Within the EMC standards, a distinction is made between _____ disturbances and _____ disturbances.

(3) There are two kinds of noise transmitting, namely _____ noise which flows into the earth, and _____ noise which flows in the circuit inside.

# Word-Building (22)　-able; -al　能…的（形容词尾）

**1. -able**　[形容词后缀]，表示"…的","能…的"

| | | | | |
|---|---|---|---|---|
| accept | v. | acceptable | adj. | 可接受的 |
| adjust | v. | adjustable | adj. | 可调节的 |
| allow | v. | allowable | adj. | 可承认的，允许的 |
| dispatch | v. | dispatchable | adj. | 可调度的 |
| fission | v.n. | fissionable | adj. | 可裂变的 |
| compare | v.n. | comparable | adj. | 可裂变的 |
| cap-（完成） | vt. | capable | adj. | 有能力的，可以…的 |
| exchange | v.n. | exchangeable | adj. | 可交换的，可替换的 |
| move | v.n. | movable | adj. | 可移动的 |
| reason | v.n. | reasonable | adj. | 合理的，有道理的 |
| rely | v. | reliable | adj. | 可靠的，可信赖的 |
| sta-（静止） | | stable | adj. | 稳定的 |
| suit | v. | suitable | adj. | 适当的，相配的 |

**2. -al; ial**　[形容词后缀]，表示"…的","有…属性的"

| | | | | |
|---|---|---|---|---|
| artifice | n. | artificial | adj. | 人造的，假的 |
| centre | n. | central | adj. | 中心的，中央的 |

# Chapter 23  Electromagnetic Field and Electromagnetic Compatibility

| | | | | |
|---|---|---|---|---|
| culture | *n.* | cultural | *adj.* | 文化的 |
| digit | *n.* | digital | *adj.* | 数字的 |
| essence | *n.* | essential | *adj.* | 本质的，基本的，精华的 |
| experiment | *n.* | experimental | *adj.* | 实验的，根据实验的 |
| function | *n.* | functional | *adj.* | 功能的 |
| industry | *n.* | industrial | *adj.* | 工业的 |
| nature | *n.* | natural | *adj.* | 自然的，自然界的 |
| part | *n.* | partial | *adj.* | 部分的，局部的 |

# Unit 7

# Review

## I. Summary of Glossaries: HVDC, FACTS, Power Quality and EMC

1. 换流器　converters
   - rectifier　　　　　　　　　　　　　　　　整流器
   - inverter　　　　　　　　　　　　　　　　逆变器
2. HVDC 分类　categories of HVDC links
   - monopolar link　　　　　　　　　　　　单极型
   - bipolar link　　　　　　　　　　　　　　双极型
   - homopolar link　　　　　　　　　　　　同极型
3. HVDC 连接　HVDC connections
   - back-to-back　　　　　　　　　　　　　背靠背的
   - point-to-point　　　　　　　　　　　　　点对点的
4. 电能质量问题　power quality problems
   - imbalance　　　　　　　　　　　　　　不平衡
   - voltage sag　　　　　　　　　　　　　　电压凹陷（暂降）
   - voltage fluctuation / flicker　　　　　　　电压波动与闪变
   - harmonic distortion　　　　　　　　　　谐波畸变
5. 相位关系　phase relationship
   - lag　　　　　　　　　　　　　　　　　领先，超前
   - lead　　　　　　　　　　　　　　　　　滞后，落后
   - be in phase with　　　　　　　　　　　与……同相
6. 电磁兼容　EMC, Electro Magnetic Compatibility
   - immunity　　　　　　　　　　　　　　抗扰度，免疫性
   - emission　　　　　　　　　　　　　　　散发，发射
   - electromagnetic environment　　　　　　电磁环境
7. 扰动的传播　urbances in EMC standards
   - radiated disturbance　　　　　　　　　　辐射的扰动
   - conducted disturbance　　　　　　　　　传导的扰动
8. 两种噪声　two kinds of noise
   - common-mode nois　　　　　　　　　　共模噪声
   - differential-mode noise　　　　　　　　差模噪声

## II. Abbreviations (Abbr.)

1. HVDC　　high voltage direct current　　　　　　　　　　　　高压直流（输电）
2. FACTS　　flexible AC transmission system　　　　　　　　　　柔性交流输电系统

| | | | |
|---|---|---|---|
| 3. | GTO | gate turn-off thyristor | 门极可关断晶闸管 |
| 4. | IGBT | insulated gate bipolar transistor | 绝缘栅型双极性晶体管 |
| 5. | IGCT | integrated gate commutation transistor | 集成门极可换向晶体管 |
| 6. | VSI | voltage source inverter | 电压源逆变器 |
| 7. | SVC | static var compensator | 静止无功补偿器 |
| 8. | SVG | static var generator | 静止无功发生器 |
| 9. | TCR | thyristor-controlled reactor | 晶闸管可控电抗器 |
| 10. | TSC | thyristor-switched capacitor | 晶闸管投切电容器 |
| 11. | TCSC | thyristor controlled series compensation | 晶闸管可控串联补偿 |
| 12. | DVR | dynamic voltage restorer | 动态电压恢复器 |
| 13. | APF | active power filters | 有源电力滤波器 |
| 14. | PWM | pulse width modulation | 脉（冲）宽（度）调制 |
| 15. | PCC | point of common coupling | 公共联结点 |
| 16. | THD | total harmonic distortion | 总谐波畸变（失真） |
| 17. | IEC | International Electrotechnical Commission | 国际电工委员会 |
| 18. | ASD | adjustable speed drive | 调速驱动器 |
| 19. | EMC | electro magnetic compatibility | 电磁兼容（性） |
| 20. | EMI | electro magnetic interference | 电磁干扰 |
| 21. | EMS | electro magnetic sensitivity | 电磁敏感性 |
| 22. | EMP | electro magnetic pulse | 电磁脉冲 |
| 23. | ESD | electric static Discharge | 静电放电 |

# Special Topic(7)　Synthesis of Words

**1. 多个单词直接组合成新词，且新词的词义就是组成它的单词的词义组合。**

（1）形如 overvoltage, underground, superconductor 的，这一类合成词也可看成是用前缀 over-, under-, super- 接上名词构成的单词。

overvoltage 过电压　　　　overload 过负荷　　　　overtime 超时
underground 地下（的）　　underwater 水下（的）　　superconductor 超导体
post-disturbance 故障后

（2）形如 superhigh 的，两个单词各自保持本义，词性不变。

superhigh 超高的　　　　bottleneck 瓶颈　　　　baseload 基本负荷

（3）形如 downwind 的，两个单词各自保持本义，组合在一起后词性发生变化。

downwind 顺风的　　　　overhead 头上的，架空的

（4）形如 full-load, no-load, off-grid, on-line 的，两个单词各自保持本义，可以当作一个单词，中间用短线连接；也可以作为两个单词，中间空格。

full-load 满负荷　　　　no-load 无载　　　　off-grid 脱（离电）网
on-line 在线　　　　　　mine-mouth 矿山口

（5）形如 tee-junction, wye-connected 的。

tee-junction T 型结点　　wye-connected 星（Y）型连接的
self-commutated 自换向的　thyristor-switched 晶闸管投切的

**2. 多个单词直接组合成新词，但新词具有特殊的词义。**

（1）形如 solid-state 的，可以根据各单词的本义按字面翻译，但合成的新词具有具体的引申义。

solid-state 字面意思：固态的。引申特指：使用电晶体的，不用真空管的

例如：Some circuit breakers are solid-state devices.

（2）fact-of-life 无法更改的事实。（可以作为一个单词，中间用短线连接；也可以作为多个单词，中间空格。）

例如：Events like voltage sags and interruptions are considered a "fact of life".

（3）state-of-the-art 达到最新技术发展水平的。（可以作为一个单词，中间用短线连接；也可以作为多个单词，中间空格。不过分开时 state of the art 意思是：技术发展水平）

例如：STATCOM is a principal state-of-the-art FACTS equipment.

（4）down-and-out 被击垮的。

（5）right of way 公用事业用地，穿越如建在铁路线、公路线、管道线设施上的一片土地。

（6）形如 back-to-back，point-to-point 的。

back-to-back 背靠背　　　　point-to-point 点对点

**3. 多个单词直接组合成新词，新词的词义需要由各单词词义适当变化得到。**

（1）frequency-dependent 独立于频率的。

例如：The speed is frequency-dependent.

（2）propeller-like 类似螺旋桨的。

**4. 单词变形之后组合成新词，新词的词义需要由各单词词义适当变化得到。**

（1）electronic-based 基于电子的。

例如：The use of power electronic-based apparatus at various voltage levels in power systems is becoming widespread due to rapid progress in power electronic technology.

（2）oil-impregnated 注入油的，浸油的。

例如：The conductors are insulated with oil-impregnated paper.

（3）coal-fired 烧煤的。

（4）flat-topped 平顶的。

例如：A wave having a crest factor less than 1.414 tends to be flat-topped.

## Knowledge and Skills (7)　　Shift of Sentence Structure

**1. 作为介词宾语的偏正名词短语转为主谓短语。**

名词短语作为表示原因等的介词的宾语时，多属这种情况。

此类介词包括 because of, for, due to, with respect to, according to，等等。

例1：Porcelain is chosen because of its resistance to deterioration when exposed to the weather, its high dielectric strength, and its ability to wash clean in rain.

选择陶瓷是因为它暴露在气象条件中时的抗老化能力、绝缘强度大以及雨中的清洗保洁能力。

说明：按照中文表述习惯，形如 its high dielectric strength 的名词短语常常译为主谓结构，即按 its dielectric strength is high 来翻译。

例2：In situations where ground currents are not tolerable or when a ground electrode is not feasible for reasons such as high earth resistivity, a third conductor is used as a metallic neutral.

在地电流不能容忍的场合或者由于地电阻率太高等原因接地电极不可行的时候，第三个导体被用作金属中性点。

说明：按照中文表述习惯，形如 high earth resistivity 的短语常常译为主谓结构，即按 earth resistivity is high 来翻译。

例3：Today's power transmission and distribution systems face increasing demands for more power, better quality with higher reliability and at lower cost as well as low environmental impact.

如今的输配电系统面临日益增长的需求，要求功率更大，因可靠性更高而质量更好，既要对环境影响小又要成本更低。

说明：介词 for 带有多个并列的宾语，这些宾语都有各自的定语，这样的名词短语在翻译时可以处理为主谓结构，译文更容易符合中文习惯。

**2. 作为动词宾语的名词短语转为主谓结构。**

具有动词含义的名词作为某些动词的宾语短语的中心词时，可以将该名词短语理解为名词+动词不定式或动名词的形式，并按主谓结构翻译。

这类动词包括 allow, permit, enable, be helpful to，等等。

例1：Once the ionized path to ground is established, the resultant low impedance to ground allows the flow of power current from the conductor to the ground and through the ground to the grounded neutral of a transformer or generator, thus completing the circuit.

接地电离通道一旦建立，所导致的对地低阻抗会使电流从导线流入大地，并经过大地流入变压器或者发电机的接地中线，因而形成一个完整回路。

说明：allows the flow of power current 相当于 allows power current to flow，而且按照后者的语序进行翻译，似乎更符合中文表述习惯。

例2：Once the ionized path to ground is established, the resultant low impedance to ground allows the flow of power current from the conductor to the ground and through the ground to the grounded neutral of a transformer or generator, thus completing the circuit.

接地电离通道一旦建立，所导致的对地低阻抗会使电流从导线流入大地，并经过大地流入变压器或者发电机的接地中线，因而形成一个完整回路。

说明：allows the flow of power current 相当于 allows power current to flow，而且按照后者的语序进行翻译，似乎更符合中文表述习惯。

**3. 介宾短语由状语转主语。**

当一个介宾短语作为状语，表示"在某方面、在某范畴"，而该方面或范畴又是主语的属性时，翻译时可以将该介宾短语转化为主语或主语的一部分。

例 1：Although insulating materials are very stable under ordinary circumstances, they may change radically in characteristics under extreme conditions of voltage stress or temperature or under the action of certain chemicals.

虽然在平常情况下绝缘材料很稳定，在极端的电压应力或温度条件下，或者在某些化学作用下，其特性会完全改变。

说明：they may change radically in characteristics 部分相当于 their characteristics may change radically，而且按照后者翻译，往往更符合中文表示习惯。

例 2：As laboratory techniques improved, different laboratories were in closer agreement on test results.

随着实验室技术的改进，不同实验室的结果近似一致。

说明：different laboratories were in closer agreement on test results 相当于 test results of different laboratories were in closer agreement，而且按照后者的语序更符合中文表述习惯。

**4. 定语从句转为新句。**

当一个作为宾语的名词带有很长的定语，或者定语从句的结构很复杂时，可以把该定语从句处理为单独的句子，这样翻译起来更通顺。不过在翻译时仍要注意相互修饰关系，尽量增补一些词语（如连词、代词等）以体现这种联系。

例 1：This provides a high degree of structural redundancy that enables the system to withstand unusual contingencies without service disruption to the consumers.

这提供了高度的结构化冗余，使系统能够承受罕见的意外事故而不会中断对用户的供电。

说明：从 that 开始到句子结束是 redundancy 的定语从句，在翻译时为了句子通顺，相关内容不必作为定语放到其所修饰的名词前面，可以处理为新的句子。

# Chapter 24
# Renewable Energy Sources and Distributed Generation

## Part 1  Importance of Renewable Energy Application

Discovering new sources of energy, obtaining an essentially inexhaustible supply of energy for the future, making energy available wherever needed, and converting energy from one form to another and using it without creating the pollution that will destroy our biosphere are among the greatest challenges facing this world today.*

The world currently relies heavily on coal, oil, and natural gas for its energy. Fossil fuels are nonrenewable, that is, they draw on finite resources that will eventually dwindle, becoming too expensive or too environmentally damaging to retrieve.* In contrast, *renewable energy resources* such as wind and solar energy are constantly replenished and will never run out.

## Part 2  Solar Energy

The sun's heat and light provide an abundant source of energy that can be harnessed in many ways. There are a variety of technologies that have been developed to take advantage of solar energy.

Sunlight, or solar energy, can be used directly for heating and lighting homes and other buildings, for generating electricity, and for hot water heating, solar cooling, and a variety of commercial and industrial uses.

- Concentrating solar power systems — Using the sun's heat to produce electricity.
- Passive solar heating and daylighting — Using solar energy to heat and light buildings.
- Photovoltaic (solar cell) systems — Producing electricity directly from sunlight.
- Solar hot water — Heating water with solar energy.
- Solar process heat and space heating and cooling — Industrial and commercial uses.

Solar power can be used in both large-scale applications and in smaller systems for the home. Businesses and industry can diversify their energy sources, improve efficiency, and save money by choosing solar technologies for heating and cooling, industrial processes, electricity, and water heating. Homeowners can also use solar technologies for heating and cooling and water heating, and may even be able to produce enough electricity to operate "off-grid" or to sell the extra electricity to the utilities, depending on local programs. The use of passive solar heating and daylighting design strategies can help both homes and commercial buildings operate more efficiently and make them more pleasant and comfortable places in which to live and work.

Beyond these localized uses of solar power, concentrating solar power systems allow power plants to produce electricity from the sun on a larger scale, which in turn allows consumers to take advantage of solar power without making the investment in personal solar technology systems.

## Part 3   Wind Power

We have been harnessing the wind's energy for hundreds of years. From old Holland to farms in the U.S., windmills have been used for pumping water or grinding grain. Today, their modern equivalent— "*wind turbines*" —can use the wind's energy to generate electricity.

Wind turbines, like windmills, are mounted on a tower to capture the most energy. At 100 feet (30 meters) or more aboveground, they can take advantage of the faster and less turbulent wind. Turbines catch the wind's energy with their propeller-like blades. Usually, two or three blades are mounted on a shaft to form a *roto*r.

A blade acts much like an airplane wing. When the wind blows, a pocket of low-pressure air forms on the downwind side of the blade. The low-pressure air pocket then pulls the blade toward it, causing the rotor to turn. This is called "*lift*". The force of the lift is much stronger than the wind's force against the front side of the blade, which is called "*drag*". The combination of lift and drag causes the rotor to spin like a propeller, and the turning shaft spins a generator to make electricity.

Wind turbines can be used as stand-alone applications, or they can be connected to a utility power grid or even be combined with a photovoltaic system. For utility-scale sources of wind energy, a large number of wind turbines are usually built close together to form a *wind plant* in a *wind farm*. Stand-alone wind turbines are typically used for water pumping or communications. However, homeowners, farmers, and ranchers in windy areas can also use wind turbines as a way to cut their electric bills.

## Part 4   Biomass Energy

The organic matter that makes up plants is known as biomass.

We have used biomass energy or "bioenergy"—the energy from plants and plant-derived materials since people began burning wood to cook food and keep warm*. Wood is still the largest biomass energy resource today, but other sources of biomass can also be used. These include food crops such as corn (for ethanol) and soybeans (for biodiesel),

# Chapter 24  Renewable Energy Sources and Distributed Generation

grassy and woody plants, residues from agriculture or forestry, and the organic component of municipal and industrial wastes such as paper mill residue and lumber mill scrap. Even the fumes from landfills (which are methane, a natural gas) can be used as a biomass energy source.

Biopower, or biomass power, is the use of biomass to generate electricity. Biomass can also be used for fuels and products that would otherwise be made from fossil fuels. In such scenarios, biomass can provide an array of benefits.

The use of biomass energy has the potential to greatly reduce greenhouse gas emissions. Carbon dioxide released by biomass is largely balanced by that captured in its growth (depending on how much energy was used to grow, harvest, and process the fuel) by photosynthesis. The use of biomass can reduce dependence on foreign oil because biofuels are the only renewable liquid transportation fuels available. Biomass energy supports agricultural and forest-product industries. The main biomass feedstocks for power are paper mill residue, lumber mill scrap, and municipal waste.

## Part 5  Hydrogen

Hydrogen is the most abundant element on the Earth. But it doesn't occur naturally as a gas. It's always combined with other elements, such as with oxygen to make water. Once separated from another element, it can be burned as a fuel or converted into electricity.

Fill vehicle fuel tanks with it instead of gasoline; pipe it to homes for heating and cooking instead of natural gas and to generate electricity onsite instead of sending electricity through transmission lines. And emit only water vapor where it is used. Fuel cells that electrochemically combine hydrogen and oxygen to produce electricity and heat offer the promise of making hydrogen an ideal universal fuel.

Most hydrogen production today is by steam-reforming natural gas. But natural gas is already a good fuel and one that is rapidly becoming scarcer and more expensive. There are, however, many possible ways to produce hydrogen with renewable energy such as thermochemical and electrolytic hydrogen, and electrochemical and biological photolytic hydrogen. Hydrogen has very high energy for its weight, but very low energy for its volume, so new technology is needed to store and transport it.* Fuel cell technology is still in early development, needing improvements in efficiency and durability.

## Part 6  Geothermal

Many technologies have been developed to take advantage of geothermal energy, the heat from the earth. This heat can be drawn from several sources: hot water or steam reservoirs deep in the earth that are accessed by drilling, geothermal reservoirs located near the earth's surface, and the shallow ground near the earth's surface that maintains a relatively constant temperature of 50°-60°F.

Geothermal energy taps the Earth's internal heat for a variety of uses. A utility can use the hot water and steam from reservoirs to drive generators and produce electricity. There are three types of geothermal power plants: *dry steam, flash steam, and binary cycle*. Other applications apply the heat produced from geothermal directly to various uses in buildings, roads, agriculture, and industrial plants. Still, others use the heat directly from the ground to provide heating and cooling in homes and other buildings. In the winter, the heat pump removes heat from the heat exchanger and pumps it into the indoor air delivery system. In the summer, the process is reversed.

In the future, geothermal resources existing miles beneath the earth's surface in the hot rock and magma may also be useful as sources of heat and energy.

## Part 7  Ocean Energy

As the world's largest solar collectors, oceans generate thermal energy from the sun. They also produce mechanical energy from the tides and waves. The gravitational pull of the moon primarily drives the tides, and the wind powers the ocean waves.

Some of the oldest ocean energy technologies use *tidal power*. All coastal areas consistently experience two high and two low tides over a period of slightly greater than 24hours. For those tidal differences to be harnessed into electricity, the difference between high and low tides must be at least 5 meters or more than 16 feet. There are only about 40 sites on the earth with tidal ranges of this magnitude.

*Wave power* devices extract energy directly from surface waves or pressure fluctuations below the surface. As believed, there is enough energy in the ocean waves to provide up to 2 terawatts of electricity. (A terawatt is equal to a trillion watts.)

A process called Ocean Thermal Energy Conversion (OTEC) uses the heat energy stored in the earth's oceans to generate electricity. OTEC works best when the temperature difference between the warmer, top layer of the ocean and the colder, deep ocean water is about 20°C (36°F).* These conditions exist in tropical coastal areas, roughly between the Tropic of Capricorn and the Tropic of Cancer.* To bring the cold water to the surface, OTEC plants require an expensive, large-diameter intake pipe, which is submerged a mile or more into the ocean's depths. As believed that OTEC could produce billions of watts of electrical power if it could become cost-competitive with conventional power technologies.

## Part 8  Distributed Energy and Production

Distributed energy refers to a variety of small, modular power-generating technologies that can be combined with load management and energy storage systems to improve the qual-

# Chapter 24  Renewable Energy Sources and Distributed Generation

ity and reliability of the electricity supply. They are "distributed" because they are placed at or near the point of energy consumption, unlike traditional "centralized" systems, where electricity is generated at a remotely located, large-scale power plant and then transmitted down power lines to the consumer.*

Distributed energy technologies can relieve transmission bottlenecks by reducing the amount of electricity that must be sent long distances down high-voltage power lines.

Implementing distributed energy can be as simple as installing a small, stand-alone electricity generator to provide backup power at an electricity consumer's site. Or it can be a more complex system, highly integrated with the electricity grid and consisting of electricity and thermal generation, energy storage, and energy management systems.

Distributed energy encompasses a wide range of technologies including wind turbines, solar power, fuel cells, microturbines, reciprocating engines, load reduction technologies, and battery storage systems. The effective use of grid-connected distributed energy resources can also require power electronic interfaces and communication and control devices for efficient dispatch and operation of generating units.

Distributed energy technologies are playing an increasingly important role in the nation's energy portfolio. They can be used to meet baseload power, peaking power, backup power, remote power, power quality, as well as cooling and heating needs.

Distributed energy also has the potential to mitigate congestion in transmission lines, reduce the impact of electricity price fluctuations, strengthen energy security, and provide greater stability to the electricity grid.

Distributed power generators are small compared with typical central-station power plants and provide unique benefits that are not available from centralized electricity generation. Many of these benefits stem from the fact that the generating units are inherently modular, which makes distributed power highly flexible. It can provide power where it is needed and when it is needed. And because they typically rely on natural gas or renewable resources, the generators can be quieter and less polluting than large power plants, which makes them suitable for on-site installation in some locations.

The use of distributed energy technologies can lead to improved efficiency and lower energy costs, particularly in combined cooling, heating, and power (CHP) applications. CHP systems provide electricity along with hot water, heat for industrial processes, space heating and cooling, refrigeration, and humidity control to improve indoor air quality.

Grid-connected distributed energy resources also support and strengthen the central-station model of electricity generation, transmission, and distribution. While the central generating plant continues to provide most of the power to the grid, the distributed resources can be used to meet the peak demands of local distribution feeder lines or major customers. Computerized control systems, typically operating over telephone lines, make it possible to operate the distributed generators as dispatchable resources, generating electricity as needed.

By siting smaller, more fuel-flexible systems near energy consumers, distributed generation avoids transmission and distribution power losses and provides a choice of energy systems to the utility customer.

Many distributed power systems produce so little noise or so few emissions that they can be located inside, or immediately adjacent to, the buildings where the power is needed. This greatly simplifies the problems of bringing power to expanding commercial, residential, and industrial areas.

Distributed power systems offer reliability for the growing number of businesses and consumers who need dependable, high-quality power to run sensitive digital equipment. They can also provide alternative, less-expensive power sources during peak price periods.

Distributed power generation technologies use a variety of fuels, including natural gas, diesel, biomass-derived fuels, fuel oil, hydrogen, sunlight, and wind.

## NEW WORDS AND EXPRESSIONS

| | | | |
|---|---|---|---|
| 1. | renewable | adj. | 可再生的，可更新的 |
| 2. | inexhaustible | adj. | 无穷尽的，用不完的 |
| 3. | biosphere | n. | 生物圈 |
| 4. | nonrenewable | adj. | 不可再生的，不可更新的 |
| 5. | dwindle | v. | 缩小 |
| 6. | retrieve | v. | 重新得到 |
| 7. | replenish | v. | 补充 |
| 8. | harness | vt. | 利用（河流等）产生动力 |
| 9. | photovoltaic | adj. | 光电的 |
| 10. | solar cell | | 太阳能电池 |
| 11. | diversify | v. | 使多样化，作多样性的投资 |
| 12. | off-grid | | 脱离电网地，独立地 |
| 13. | concentrating | adj. | 集中的，浓缩的 |
| 14. | windmill | n. | 风车，风车房 |
| 15. | grind | v. | 磨（碎），碾（碎） |
| 16. | turbulent | adj. | 狂暴的，动荡的 |
| 17. | propeller | n. | 螺旋桨 |
| 18. | blade | n. | 桨叶，叶片 |
| 19. | shaft | n. | 轴，杆状物 |
| 20. | downwind | adv. | 顺风地 |
| 21. | air pocket | | 气陷，气穴 |
| 22. | lift | n. | 空气浮力 |
| 23. | rancher | n. | 牧场（或农场）主 |
| 24. | biomass | | 生物质，用作能源的植物材料 |
| 25. | organic matter | | 有机物 |
| 26. | bioenergy | n. | 生物能 |
| 27. | ethanol | n. | 乙醇，酒精 |
| 28. | biodiesel | n. | 生物柴油 |
| 29. | residue | n. | 剩余物 |

# Chapter 24  Renewable Energy Sources and Distributed Generation

| 30. | municipal | adj. | 市政的 |
| --- | --- | --- | --- |
| 31. | lumber | n. | 木材 |
| 32. | fume | n. | （浓烈或难闻的）烟 |
| 33. | landfill | n. | 垃圾掩埋 |
| 34. | methane | n. | 甲烷，沼气 |
| 35. | scenario | n. | 特定情节 |
| 36. | greenhouse | n. | 温室，花房 |
| 37. | photosynthesis | n. | 光合作用 |
| 38. | feedstock | n. | 给料，原料 |
| 39. | hydrogen | n. | 氢 |
| 40. | scarcer | adj. | 缺乏，不足（scarce 的比较级） |
| 41. | electrolytic | adj. | 电解的，由电解产生的 |
| 42. | photolytic | adj. | 光解的 |
| 43. | durability | n. | 耐用性，耐久力 |
| 44. | geothermal | adj. | 地热的，地温的 |
| 45. | drilling | n. | 钻井 |
| 46. | tap | vt. | 使流出 |
| 47. | binary cycle | n. | 双汽循环 |
| 48. | heat pump | n. | 热（力）泵，蒸汽泵 |
| 49. | magma | n. | 岩浆 |
| 50. | tide | n. | 潮汐 |
| 51. | gravitational | adj. | 重力的 |
| 52. | trillion | num. | 万亿 |
| 53. | tropic | n. | 回归线，热带 |
| 54. | Capricorn | n. | 摩羯宫，摩羯（星）座 |
| 55. | Cancer | n. | 巨蟹星座 |
| 56. | modular | adj. | 模块化的，有标准组件的 |
| 57. | bottleneck | n. | 瓶颈 |
| 58. | encompass | vt. | 包含或包括 |
| 59. | reciprocating | n. | 往复式发动机 |
| 60. | portfolio | n. | 投资组合 |
| 61. | congestion | n. | 拥塞 |
| 62. | central-station | n. | 总（电）站，总厂，中心电站 |
| 63. | stem | v. | 滋生 |
| 64. | humidity | n. | 湿气，潮湿，湿度 |
| 65. | diesel | n. | 柴油 |

## PHRASES

**1. run out: 被用完**

Example: In contrast, renewable energy resources such as wind and solar energy are constantly replenished and will never <u>run out</u>.

相比之下，风能和太阳能多可再生，能源不断补充，永远不会耗尽。

2. take advantage of: 利用    a variety of: 多种的

Example: There are a variety of technologies that have been developed to take advantage of solar energy.

多种技术开发出来利用太阳能。

## COMPLICATED SENTENCES

**1. Discovering new sources of energy, obtaining an essentially inexhaustible supply of energy for the future, making energy available wherever needed, and converting energy from one form to another and using it without creating the pollution that will destroy our biosphere are among the greatest challenges facing this world today.**

译文：发现新能源，得以在将来提供基本上用之不竭的能量供应，使能源随用随有，并且把能量从一种形式转换为另一种形式以及使用时不产生会破坏我们生物圈的污染，这是如今世界所面临的最大挑战之一。

说明：句中 available wherever needed 意思是"哪里需要哪里就可以得到"。词组 convert ... from ... to 表示"由…转换为…"。among 表示"在…中，是…中之一"。

**2. Fossil fuels are nonrenewable, that is, they draw on finite resources that will eventually dwindle, becoming too expensive or too environmentally damaging to retrieve.**

译文：化石燃料是不可再生的，也就是说，它们利用的是最终会减少、变得过于昂贵或太破坏环境而无法重新获取的有限资源。

说明：词组 that is 表示"即，就是，也就是说"。词组 too ... to ... 表示"太……以至不能"。

**3. We have used biomass energy or " bioenergy "    —the energy from plants and plant-derived materials since people began burning wood to cook food and keep warm.**

译文：自从人们开始烧柴来做饭和取暖，我们就在使用生物质能或"生物能源"——来自于植物或其提取物的能量。

**4. Hydrogen has very high energy for its weight, but very low energy for its volume, so new technology is needed to store and transport it.**

译文：氢相对其重量而言能量很高，但相对其体积则能量很低，因此需要新的技术来存储和输送它。

说明：for 在此表示"对应于，相对于"。

**5. OTEC works best when the temperature difference between the warmer, top layer of the ocean and the colder, deep ocean water is about 20°C (36°F).**

译文：当温暖的海洋表层和较冷的深层海水之间温差约为摄氏20度（华氏36度）时，OTEC效果最好。

说明：不要被句中的逗号迷惑，逗号所分隔的是同一主体的多个定语。

# Chapter 24  Renewable Energy Sources and Distributed Generation

**6. These conditions exist in tropical coastal areas, roughly between the Tropic of Capricorn and the Tropic of Cancer.**

译文：这些情况大概出现在介于南回归线和北回归线之间的热带海岸区。

说明：roughly between 引领的短语作为 tropical coastal areas 的状语。Tropic 是"回归线"，Capricorn 指的是"摩羯星座"，Cancer 指的是"巨蟹星座"，基于两个星座在天球上的位置，Tropic of Capricorn 和 the Tropic of Cancer 分别对应"南回归线"和"北回归线"。

**7. They are "distributed" because they are placed at or near the point of energy consumption, unlike traditional "centralized" systems, where electricity is generated at a remotely located, large-scale power plant and then transmitted down power lines to the consumer.**

译文：它们是"分布式的"，因为它们位于或接近能量消耗的地点，不像传统的"集中"系统，电能在位置遥远的大规模电厂产生，然后沿着电力线输送给消费者。

说明：where 引导的定语从句修饰"centralized" systems。介词 down 在此表示"沿着，顺着"。placed at or near the point 部分有省略，补全应为 placed at the point or near the point。

## ABBREVIATIONS (ABBR.)

| | | | |
|---|---|---|---|
| 1. | OTEC | ocean thermal energy conversion | 海洋热能转换系统 |
| 2. | CHP | cooling, heating, and power | 冷却、加热和供电 |

## SUMMARY OF GLOSSARY

1. 负荷功率　load power
   baseload power　　　　　　　　　基荷功率，基本负荷功率
   peaking power　　　　　　　　　峰荷功率
   backup power　　　　　　　　　　备用功率
2. 海洋能　ocean energy
   tidal power generation　　　　　　潮汐发电
   wave power generation　　　　　海浪发电，浪力发电
   thermal energy conversion　　　　（海洋）热能转换
3. 制氢方法　method of Hydrogen production
   electrochemical　　　　　　　　　电化学的
   thermochemical　　　　　　　　　热化学的
   electrolytic　　　　　　　　　　　电解的
   photolytic　　　　　　　　　　　　光解的

# EXERCISES

**1. Translate the following words or expressions into Chinese.**

(1) greenhouse air　　　　(2) biomass　　　　(3) geothermal energy

(4) solar cell　　　　(5) wind power　　　　(6) heat pump

**2. Translate the following words or expressions into English.**

(1) 可再生能源　　　(2) 瓶颈　　　　　　(3) 分布式发电
(4) 潮汐　　　　　　(5) 太阳能　　　　　(6) 燃料电池

**3. Fill in the blanks with proper words or expressions.**

(1) Discovering new sources of energy, obtaining an essentially _____ supply of energy for the future, making energy available wherever _____, and converting energy from one form to another and using it without creating the _____ that will destroy our biosphere are among the greatest challenges facing this world today.

(2) The sun's heat and light provide an abundant source of energy that is called _____ energy. Wind _____ can use the wind's energy to generate electricity. The organic matter that makes up plants is known as _____, the energy derived from them is called _____. _____ is the most abundant element on the Earth; once separated from another element, it can be burned as a fuel or converted into electricity. _____ energy is the heat from the earth.

(3) As the world's largest solar collectors, oceans generate _____ energy from the sun. They also produce _____ energy from the _____ which is driven by the gravitational pull of the moon and _____ which is powered by the wind.

(4) _____ energy refers to a variety of small, modular power-generating technologies that can be combined with _____ management and _____ storage systems to improve the quality and reliability of the electricity supply.

# Word-Building (23)　photo-; bio-　电，磁

### 1. photo-　[前缀] 表示：光，光电

| | | |
|---|---|---|
| photocell | n. | 光电池 |
| photoconduction | n. | 光电导（性） |
| photochopper | n. | 光线断路器 |
| photocoupler | n. | 光电耦合器 |
| photoreceptor | n. | 光感受器，感光器 |
| photocon | n. | 光（电）导元件，光导器件 |
| photod | n. | 光电二极管 |
| photoeffect | n. | 光电效应 |
| photovoltaic | adj. | 光电的 |
| photochemical | adj. | 光化学的 |
| photolytic | adj. | 光解的 |

### 2. bio-　[前缀] 表示：生物，生命

| | | |
|---|---|---|
| biology | n. | 生物，生物学 |

# Chapter 24　Renewable Energy Sources and Distributed Generation

| | | |
|---|---|---|
| biomass | *n.* | 生物质 |
| bioenergy | *n.* | 生物能 |
| biopower | *n.* | 生物电源 |
| biofuel | *n.* | 生物燃料 |
| biodiesel | *n.* | 生物柴油 |
| bioelectric | *adj.* | 生物电的 |
| biological | *adj.* | 生物学的，生物的 |

# Chapter 25
# Multi-energy Utilization

## Part 1　Importance of Multi-energy Utilization

There are significant efforts worldwide at multiple levels, from research to policy initiatives, to support the integration of renewable electricity resources into the power system, particularly by deploying innovative concepts such as the Smart Grid. However, meeting challenging environmental targets and guaranteeing secure and affordable energy to present and future generations require clear strategies addressing all energy sectors, and not only electricity.*

MES (multi-energy systems) whereby electricity, heat, cooling, fuels, transport, and so on optimally interact with each other at various levels (for instance, within a district, city, or region) represent an important opportunity to increase technical, economic and environmental performance relative to "classical" energy systems whose sectors are treated "separately" or "independently". * This performance improvement can take place at both the operational and the planning stage.

## Part 2　Energy Internet

Energy Internet is a complex multi-network flow system with a power system as the core, Internet and other cutting-edge information technologies as the foundation, distributed renewable energy as the main primary energy, and closely coupled with natural gas network, transportation network, and other systems.

Energy Internet is composed of four complex network systems, namely the power system, transportation system, natural gas network, and information network, which are closely coupled. It draws on the concept of the Internet, builds energy infrastructure <u>from the bottom up</u>, and increases the flexible access and local consumption of distributed renewable energy through the open peer-to-peer interconnection of autonomous energy units like microgrids. *

The specific differences between energy Internet and smart grid are as follows:

(1) The physical entity of the smart grid is mainly a power system, while the physical entity of the energy Internet is composed of a power system, transportation system, and natural gas network.

(2) In a smart grid, energy can only be transmitted and used in the form of electric ener-

# Chapter 25 Multi-energy Utilization

gy; In the energy Internet, energy can be converted into electric energy, chemical energy, heat energy, and other forms.

(3) At present, the research on smart grids mainly adopts local absorption and control for distributed devices such as distributed power generation, energy storage and, controllable load. In the energy Internet, the research focus will shift from local consumption to <u>wide-area coordination</u>.

(4) The information system of the smart power grid <u>is dominated by</u> the traditional industrial control system, and in the energy Internet, open information networks such as the Internet will play a greater role.

## Part 3   Ubiquitous Electric Internet of Things

"Ubiquitous electric Internet of Things" means that with the support of related technologies of the Internet of Things, the people and things involved in the power system are integrated and connected to realize the sharing of common data and serve the related structure of the whole power system. In this way, the integration of business flow, information flow, and energy flow can be realized in each business link of the power grid and at different levels of voltage, so as to improve the security of the system and the efficiency of operation.

The Ubiquitous electric Internet of Things comes from the smart grid. The traditional power generation, power transformation, transmission, power supply, and power dispatching within the power system, as well as the grid connection and intelligent management and control of the power supply system in the field of new energy, are called the Intranet of the ubiquitous electric Internet of Things. With the development of intelligent power management technology for electrical equipment, the electric Internet of Things will inevitably <u>spill over</u> to the power end, making the existing Internet of Things enter the era of intelligent power management. The part of the existing Internet of Things that spills over to the field is called the external network.

## Part 4   Integrated Energy System

As the physical basis of energy Internet, an Integrated Energy System (IES), is an integrated energy production and marketing system that organically coordinates energy generation and conversion, transmission and distribution, storage and consumption in the process of planning, construction, and operation. It is a typical form of energy system covering various links of "source - network - load - storage".

The system architecture of the comprehensive energy system can be divided into two types: the first type, which takes electric energy as the core and uses equipment such as Combined Cooling Heating and Power (CCHP) to couple the heat and gas network to realize "electric energy replacement". And through the multi-energy optimization configuration platform, the unified connection of cold, heat, electricity source channels of the second category. Due to the advantages of convenient construction, high energy conversion efficiency, and reliable power supply, the first type of system is more widely used.

The purpose of comprehensive energy system construction can be divided into two aspects: energy supply, to achieve the coordination and optimization of multiple energy sources, improve energy conversion efficiency, ensure energy security, and promote the consumption of clean energy; in terms of energy use, energy demand should be coordinated, energy use should become cascade utilization according to energy quality and load categories, and energy consumption and greenhouse gas emissions should be reduced in the supply process.

## Part 5    Energy Hub

The model of an energy hub, which reflects the static conversion and storage between energy systems, was first proposed by a research team at ETH Zurich in Switzerland. The concept comes from the very simple idea that no matter how complex the coupling relationship between electricity, heat, and gas in a multi-energy system is, it requires the input of various forms of energy to be converted into other forms of energy as the output of the system. In the abstract, a multi-energy system can be summarized as the input-output two-port network as shown in the figure 25.1. The box in the middle is the multi-energy system to be analyzed, namely the energy hub. The use of an energy hub to model a multi-energy system and realize the collaborative optimization of various forms of energy will fully consider the mutual benefit and complementarity of various forms of energy and produce some direct synergistic effects.

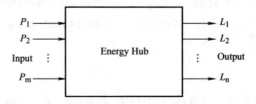

**Figure 25.1**    Energy hub

### NEW WORDS AND EXPRESSIONS

1.　initiative　　　　　　　　　　　*n.*　　　措施，倡议
2.　couple　　　　　　　　　　　　*v.*　　　连接，成对，耦合
3.　micro-grid　　　　　　　　　　　　　　微网
4.　physical entity　　　　　　　　　　　　物理实体
5.　wide-area coordination　　　　　　　　广域协调

# Chapter 25  Multi-energy Utilization

| | | | |
|---|---|---|---|
| 6. | smart grid | | 智能电网 |
| 7. | convert | v. | （使）转变，（使）转换 |
| 8. | absorption | n. | 吸收，吸纳，同化 |
| 9. | power dispatching | | 配电 |
| 10. | intranet | n. | 内网 |
| 11. | collaborative | adj. | 合作的，协作的 |
| 12. | mutual | adj. | 相互的，彼此的，共同的，共有的 |
| 13. | synergistic | adj. | 协同的，协作的，协同作用的 |

## PHRASES

**1. from the bottom up: 自底向上**

Example: It draws on the concept of the Internet, builds energy infrastructure from the bottom up.

它借鉴了互联网的概念，自底向上建设能源基础架构。

**2. be dominated by: 由……主导**

Example: The information system of the smart power grid is dominated by the traditional industrial control system, and in the energy Internet, open information networks such as the Internet will play a greater vole.

**3. spill over: 溢出，外溢**

Example: With the development of intelligent power management technology for electrical equipment, the electric Internet of Things will inevitably spill over to the power end.

随着用电设备的智能化电源管理技术发展，电力物联网不可避免地会向用电端外溢。

**4. be summarized as: 被概括**

Example: In the abstract, a multi-energy system can be summarized as the input-output two-port network as shown in the figure 25.1.

多能源系统能被抽象简化为如图 25.1 所示的输入输出二端网络。

## COMPLICATED SENTENCES

**1. However, meeting challenging environmental targets and guaranteeing secure and affordable energy to present and future generations require clear strategies addressing all energy sectors, and not only electricity.**

译文：然而，要实现这一具有挑战性的环境目标，并能为今世后代获得安全和充裕的能源，需要做出针对电力以及所有能源部门的明确战略。

说明：meeting challenging...targets 和 guaranteeing...energy 是用现在分词短语做主语的表达方式，address 在这里的意思是设法解决。

**2. MES (multi-energy systems) whereby electricity, heat, cooling, fuels, transport, and so on optimally interact with each other at various levels (for instance, within a district, city or region) represent an important opportunity to increase technical, economic and environmental performance relative to "classical" energy systems whose sectors are treated "separately" or "independently".**

译文：与单独或者说分立管理的传统能源系统相比，MES（多能源系统）这一保证电力、热力、冷能、燃料、运输等在不同层次（例如在一个地区、城市或区域内）实现最佳相互作用的系统，是提高技术、经济和环境水平的重要途径。

说明：本句 interact with each other 表示相互作用，单词加双引号的意思有表示引语、特定称谓、特殊含义及强调，本句"separately"和"independently"两处的双引号均表示强调，小括号中的内容表示附加或补充，本句中两处小括号的使用均表示补充说明。为了便于理解，这个句子可以拆解为：

Relative to "classical" energy systems, MES represents an important opportunity to increase technical, economic and environmental performance; The sectors of classical energy systems are treated "separately" or "independently"; The sectors of MES interact with each other at various levels, for instance, within a district, city or region; The sectors of MES include electricity, heat, cooling, fuels, transport and so on.

**3. It draws on the concept of the Internet, builds energy infrastructure from the bottom up, and increases the flexible access and local consumption of distributed renewable energy through the open peer-to-peer interconnection of autonomous energy units like microgrids.**

译文：它借鉴了互联网理念，自底向上构建能源基础设施，通过微网等类似能量自治单元的开放对等互联，增加分布式可再生能源的灵活接入和就地消纳。

说明：本句的主结构 It draws on..., builds..., and increases the flexible access and local consumption。短语 draws on 在这里表示借鉴，from the bottom up 在这里表示自底向上或自下而上，open 在这里是形容词，表示开放的。

## ABBREVIATIONS (ABBR.)

| | | | |
|---|---|---|---|
| 1. | MES | multi-energy systems | 多能源系统 |
| 2. | IES | integrated energy system | 综合能源系统 |
| 3. | CCHP | combined cooling heating and power | 冷热电联供 |

## SUMMARY OF GLOSSARY

1. 多能源系统种类　types of Multi-energy system
   Ubiquitous electric Internet of Things　　　泛在电力物联网
   multi-network flow system　　　　　　　　多能流网络

# Chapter 25  Multi-energy Utilization

**2.**　　能源利用方式　　methods of energy use

　　　　primary energy　　　　　　　　　　　　　一次能源
　　　　source - network - load - storage　　　　源 – 网 – 荷 – 储
　　　　carried out in steps　　　　　　　　　　梯级利用
　　　　peer-to-peer interconnection　　　　　　对等互联

# EXERCISES

**1. Translate the following words or expressions into Chinese.**

(1) couple　　　　　(2) absorption　　　　　(3) initiative
(4) convert　　　　(5) wide-area coordination　(6) mutual
(7) synergistic　　　(8) collaborative

**2. Translate the following words or expressions into English.**

(1) 智能电网　　　　(2) 能源互联网　　　　(3) 泛在电力物联网
(4) 综合能源系统　　(5) 能量枢纽　　　　　(6) 耦合
(7) 分布式发电　　　(8) 多能流系统　　　　(9) 冷热电联产

**3. Fill in the blanks with proper words or expressions.**

(1) Energy Internet is composed of four complex network systems, namely _____, _____, _____, and _____, which are closely coupled.

(2) And through the multi-energy _____, the unified connection of cold, heat, electricity source channels of the second category.

(3) The part of the existing Internet of Things that spills over to the field is called _____.

# Word-Building (24)　trans-　越过、超；转移

**1. trans-　前缀，表示：越过，横过，超**

| | | | |
|---|---|---|---|
| national | transnational | *adj.* | 超越国界的 |
| normal | transnormal | *adj.* | 超出常规的 |
| personal | transpersonal | *adj.* | 超越个人的 |
| marine | transmarine | *adj.* | 海外的 |
| frontier | transfrontier | *adj.* | 在国境外的 |

**2. trans-　前缀，表示：转移，变换**

| | | | |
|---|---|---|---|
| form | transform | *v.* | 改变 |
| position | transposition | *v.* | 调换 |

| | | | |
|---|---|---|---|
| code | transcode | v. | 译密码，转换代码 |
| migrate | transmigrate | v. | 移居 |
| plant | transplant | v. | 移植 |
| ship | transship | v. | 转船（车） |
| port | transport | v. | 运输 |
| late | translate | v. | 翻译 |

# Chapter 26
# Energy Storage System

## Part 1  Concept and Classification of Energy Storage

In recent years, the energy demand of human society has increased, while relatively environmentally sustainable renewable energy has developed rapidly, thus highlighting the importance of energy storage technology. The essence of energy storage technology is to store excess energy and release it when needed. Different energy storage systems can be selected for different occasions and needs. Energy storage systems can be divided into the following categories according to the principle of energy storage technology and differences in storage forms.

(1) Electric energy storage: including supercapacitor and super-conducting magnetic energy storage, etc.

(2) Mechanical energy storage: including flywheel energy storage, pumped storage, compressed air energy storage, etc.

(3) Chemical energy storage: it can be subdivided into electrochemical energy storage, chemical energy storage, and thermochemical energy storage. Electrochemical energy storage includes lead acid, nickel metal hydride, lithium-ion and other conventional batteries and zinc-bromine, all vanadium REDOX, and other liquid flow batteries. Chemical energy storage includes fuel cells and metal-air cells. Thermochemical energy storage includes hydrogen storage by solar energy and the use of solar dissociation and recombination of ammonia or methane.

(4) Thermal energy storage: including aquifer energy storage system, liquid air energy storage, sensible heat energy storage and latent heat storage, and other high-temperature energy storage.

## Part 2  Commonly Used Energy Storage Technology

The basic principles and characteristics of several energy storage technologies are described as follows:

• Pumped storage is the most widely used and most mature technology of large-scale energy storage technology. The working process is as follows: when the power demand is low, the electricity is used to pump the water from the lower reservoir to the upper reservoir, and the electricity is converted into potential energy storage. When the electricity demand is high, the water in the upper reservoir can be released and returned to the lower reservoir to drive the turbine to generate electricity, thus realizing the conversion of potential energy to electric energy. The structure of the pumped-storage station is shown in Figure 26.1.

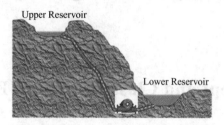

**Figure 26.1** The structure of the pumped-storage station

• Flywheel energy storage: a kind of mechanical energy storage. The electric energy can be converted into rotating kinetic energy for storage, which is mainly composed of motor, bearing, power electronic components, rotating body, and shell, as shown in Figure 26.2. When storing energy, the motor drives the flywheel to rotate at high speed, and when releasing energy, the flywheel drives the generator to generate electricity.

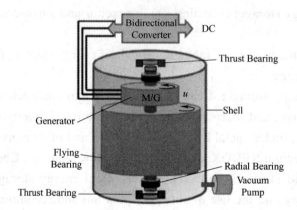

**Figure 26.2** The structure of the flywheel

• Battery Energy Storage System (BESS): consists of a battery, inverter, battery energy storage system control devices, auxiliary equipment (safety, environmental protection equipment), and other parts. The structure of the battery includes positive and negative poles (also known as anode and cathode when charging), separator, electrolyte, shell, etc.

• Superconducting energy storage: can store electric energy in the form of direct current in the ring inductor made of superconducting materials, almost realizing zero current loss. It essentially stores magnetic energy in the form of direct current, using coils of superconductors with almost zero resistance. When energy needs to be stored, alternating current is converted to direct current and stored in the coil. When electric energy needs to be released, it is converted to alternating current and fed back to the grid. In order to maintain the super-

# Chapter 26　Energy Storage System

conducting state with zero resistance, a 77K refrigeration environment is generally achieved through liquid nitrogen refrigeration.

• Compressed air energy storage: a technology based on the development of gas turbine energy storage technology, the stored energy in the form of compressed air. The following diagram shows the basic structure of a compressed air energy storage system. When there is a surplus of electricity, the compressor is driven by electricity to compress the air and store it in the chamber; When electricity is needed, high-pressure air in the chamber is released to drive a generator to produce electricity.

• Supercapacitor energy storage: the essence of which is a large capacitance, in the form of the electric field can store energy. It can be divided into a double-layer capacitor and a Faraday quasi-capacitor (also called a pseudo-capacitor).

## Part 3　The Role of Energy Storage Technology in Power System

Energy storage technology can effectively alleviate the intermittency and random volatility of renewable energy generation, improve power quality, and strengthen the deployment elasticity and toughness of the power grid. In the system, an energy storage device plays a very important role, its role is mainly manifested in the following aspects.

• Improve security and stability of the system operation, including Peak-cutting and valley-filling. Increase reserve and reduce the impact of large disturbances. Power support improves the system's energy supply stability.

• Implement the coupling between different energy networks, improve the flexibility of multiple energy systems and collaborate.

• Make full use of renewable energy: support a high proportion of renewable energy access. Reduce waste of energy such as wind and light.

• Guarantee the quality of power supply: smooth power fluctuations, guarantee the quality of the power supply.

• Ensure the continuity of power supply: as an emergency, to solve the instantaneous energy supply shortages or a short time. As a backup, to solve the problem of chronic energy supply.

• Support new energy management mode.

• Improve the controllability of distributed systems.

## Part 4　Electric Vehicle

An electric vehicle (EV) is a vehicle that is powered by On-board power and uses a battery as its energy storage unit. When idle, charge the battery fully, and when traveling, use the energy in the battery as a source of power.

A battery management system (BMS) is a system for the management and control of power batteries. Its core function is to estimate and detect the status of batteries, especially the State of charge (SOC). Battery SOC, on the other hand, reflects the remaining power in the battery. The accuracy of estimating battery SOC directly affects the vehicle's core performance and safety functions such as driving range and maximum output power.*

The relationship between electric vehicles and the power grid can be divided into two aspects:

The concept of V2G is proposed based on the energy storage capacity of electric vehicles. Its core idea is to use the energy storage of a large number of electric vehicles as a buffer between the power grid and renewable energy. In other words, as the battery itself is an energy storage device, it can provide feedback when the power grid needs energy.* This mode is Vehicle to Grid (V2G).

Electric vehicles should use charging piles, whose input is directly connected to the AC power grid, and whose output is equipped with a charging plug for charging electric vehicles. Therefore, the process of charging electric vehicles by charging piles is essentially a process of storing electric energy from the grid.

## NEW WORDS AND EXPRESSIONS

| | | | |
|---|---|---|---|
| 1. | essence | n. | 本质，要素 |
| 2. | supercapacitor | n. | 超级电容器 |
| 3. | flywheel | n. | 飞轮 |
| 4. | thermochemical | adj. | 热化学的 |
| 5. | conventional | adj. | 传统的，惯用的 |
| 6. | dissociation | n. | 分解，分离，分裂 |
| 7. | recombination | n. | 复合，重组 |
| 8. | aquifer | n. | 地下水层，渗透性含水石层 |
| 9. | chamber | n. | 腔室 |
| 10. | pseudocapacitor | n. | 准电容器，赝电容器 |
| 11. | intermittency | n. | 间歇性，间歇现象 |
| 12. | implement | v. | 执行，贯彻 |
| 13. | feedback | n. | 回馈 |

## PHRASES

**1. be subdivided into: 被再分成**

Example: It can be subdivided into electrochemical energy storage, chemical energy

# Chapter 26　Energy Storage System

storage and thermochemical energy storage.

它可细分为电化学储能、化学储能以及热化学储能等。

## COMPLICATED SENTENCES

**1. Superconducting energy storage can store electric energy in the form of direct current in the ring inductor made of superconducting materials, almost realizing zero current loss.**

译文：超导磁储能可将电能以直流电流的形式存储于由超导材料制成的环形电感器中，几乎实现电流零损耗。

说明：in the form of direct current 用来修饰 electric energy，而 made of superconducting materials 则用来修饰 ring inductor，这个句子可以拆解为：

Superconducting energy storage can store electric energy; The electric energy is stored in the form of direct current; The electric energy is stored in the ring inductor; The ring inductor is made of superconducting materials; Superconducting energy storage can almost realize zero current loss.

**2. The accuracy of estimating battery SOC directly affects the vehicle's core performance and safety functions such as driving range and maximum output power.**

译文：对电池 SOC 估计的准确性直接影响了整车续航里程、可输出最大功率等车辆核心性能和安全功能。

说明：本句的主结构为 The accuracy ... affects ... and safety functions...。Such as 引导的从句用来对 safety functions 进行补充解释。

**3. In other words, as the battery itself is an energy storage device, it can provide feedback when the power grid needs energy.**

译文：换句话说，由于电池本身是储能设备，所以可以在电网需要能量时进行反馈。

说明：本句中 as 表示由于，when 引导时间定语用来修饰 provide feedback。

## ABBREVIATIONS (ABBR.)

| | | | |
|---|---|---|---|
| 1. | BESS | battery energy storage system | 电池能量管理系统 |
| 2. | EV | electric vehicle | 电动汽车 |
| 3. | BMS | battery management system | 电池管理系统 |
| 4. | SOC | state of charge | 智能续航系统 |
| 5. | V2G | vehicle to grid | 电动汽车反馈系统 |

## SUMMARY OF GLOSSARY

1. **Energy storage　储能**
   super-conducting magnetic energy storage　　　　　　　　　　　超导磁储能

| | |
|---|---|
| flywheel energy storage | 飞轮储能 |
| pumped storage | 抽水蓄能储能 |
| compressed air energy storage | 压缩空气储能 |
| thermochemical energy storage | 热化学储能 |
| Lead Acid Storage Batteries | 铅酸蓄电池 |
| nickel metal hydride | 金属氢化镍 |
| lithium-ion battery | 锂离子电池 |
| zinc-bromine battery | 锌溴电池 |
| Vanadium Redox Battery | 钒电池 |
| sensible heat energy storage | 显热储能 |
| latent heat storage | 潜热储能 |
| double layer capacitor | 双层电容器（储能） |
| Faraday quasi-capacitor | 法拉第电容器 |
| Peak-cutting and valley-filling | 削峰填谷 |

2. **Energy Vehicle** 电动汽车

| | |
|---|---|
| On-board power | 车载电源 |
| charging pile | 充电桩 |
| charging plug | 充电插头 |

# EXERCISES

**1. Translate the following words or expressions into Chinese.**

(1) essence　　　　　(2) solar dissociation　　　(3) lower reservoir
(4) bearing　　　　　(5) auxiliary　　　　　　　(6) in the form of
(7) intermittency　　 (8) Peak-cutting　　　　　(9) valley-filling
(10) as a backup　　 (11) charging pile

**2. Translate the following words or expressions into English.**

(1) 飞轮　　　　　　(2) 腔室　　　　　　　　　(3) 轴承
(4) 线圈　　　　　　(5) 图解　　　　　　　　　(6) 超导环境
(7) 零电流损耗　　　(8) 充电桩　　　　　　　　(9) 车载电源

**3. Fill in the blanks with proper words or expressions.**

(1) Mechanical energy storage: including flywheel energy storage, _____, _____, etc.

(2) The structure of the battery includes positive and negative poles (also known as _____ and _____ when charging), separator, electrolyte, shell, etc.

(3) In other words, as the battery itself is an energy storage device, it can provide feedback when the power grid needs energy. This mode is _____.

# Chapter 26  Energy Storage System

## Word-Building (25)　dia-;per-　通过，遍及

**1. dia-**　前缀，表示：通过，横过

| | | | |
|---|---|---|---|
| meter | diameter | n. | 通径 |
| gram | diagram | n. | 通过图来解释（图解） |

**2. per-**　前缀，表示：贯穿、全、自始至终

| | | | |
|---|---|---|---|
| spective | perspective | n. | 透视 |
| manent | permanent | adj. | 永久的 |
| vade | pervade | v. | 遍及 |
| ambulate | perambulate | v. | 步行巡视、闲逛 |

# Unit 8

# Review

## I. Summary of Glossaries

1. 负荷功率　load power
   - baseload power　　　　　　　　　　　基本负荷功率
   - peaking power　　　　　　　　　　　　峰值功率
   - backup power　　　　　　　　　　　　备用功率
2. 多能源系统种类　types of Multi-energy system
   - Ubiquitous electric Internet of Things　　泛在电力物联网
   - multi-network flow system　　　　　　多能流网络
3. 储能　energy storage
   - super-conducting magnetic energy storage　超导磁储能
   - flywheel energy storage　　　　　　　飞轮储能

## II. Abbreviations (Abbr.)

| | | | |
|---|---|---|---|
| 1. | OTEC | ocean thermal energy conversion | 海洋热能转换系统 |
| 2. | CHP | cooling, heating, and power | 冷却、加热和供电 |
| 3. | DG | distributed generation | 分布式电源 |
| 4. | MES | multi-energy systems | 多能源系统 |
| 5. | IES | integrated energy system | 综合能源系统 |
| 6. | CCHP | combined cooling heating and power | 冷热电联供 |
| 7. | BESS | battery energy storage system | 电池能量管理系统 |
| 8. | EV | electric vehicle | 电动汽车 |
| 9. | BMS | battery management system | 电池管理系统 |
| 10. | SOC | state of charge | 智能续航系统 |

## 参考文献

[1] YANG R H, JIN J X, ZHOU Q, et.al. Superconducting Magnetic Energy Storage Based DC Unified Power Quality Conditioner with Advanced Dual Control for DC-DFIG[J]. Journal of Modern Power Systems and Clean Energy, 2022, 10(5): 1385-1400.

[2] 桑晓明, 刘会. 新能源汽车动力电池生产企业的安全文化体系构建: 评《电动汽车动力电池系统安全分析与设计》[J]. 电池, 2022, 52(2): 238-239.

[3] 何立民. 从智能电网、物联网到泛在电力物联网 [J]. 单片机与嵌入式系统应用, 2022, 22(4): 3-5; 10.

[4] 邱锡鑫. 泛在电力物联网的数据挖掘与智能处理技术研究 [D]. 华东师范大学, 2021.

[5] 吴皓文, 王军, 龚迎莉, 等. 储能技术发展现状及应用前景分析 [J]. 电力学报, 2021, 36(5): 434-443.

[6] 余双. 考虑 P2G、需求响应和激励政策的 CCHP 系统智能优化调度 [D]. 宁夏大学, 2021.

[7] 朱永强, 郝嘉诚, 赵娜, 等. 能源互联网中的储能需求、储能的功能和作用方式 [J]. 电工电能新技术, 2018, 37(2): 68-75.

[8] 贾宏杰, 王丹, 徐宪东, 等. 区域综合能源系统若干问题研究 [J]. 电力系统自动化, 2015, 39(7): 198-207.

[9] 董朝阳, 赵俊华, 文福拴, 等. 从智能电网到能源互联网: 基本概念与研究框架 [J]. 电力系统自动化, 2014, 38(15): 1-11.